DEPARTMENT OF THE ENVIRONMENT

Waste Management Paper No 28

Recycling

A memorandum providing guidance to local authorities on recycling

London : HMSO

Recycled Paper

It is intended to further revise and update this Waste Management Paper in the future. Any suggestions or enquiries should be addressed to:

Department of the Environment
Wastes Technical Division
Romney House
43 Marsham Street
London SW1P 3PY

WASTE MANAGEMENT PAPER No 28

Contents

Summary

The Government has set a target of recycling half of the country's recyclable household waste, which is estimated to equal about 25% of all household waste. Local authorities have a key role to play in achieving this target, because of their responsibilities for collection and disposal of household waste.

The Environmental Protection Act 1990 introduces a waste management framework which is designed to increase recycling levels. One of the main principles underlying Part II of the Act ('Waste on Land') is that recycling should be an integral part of authorities' waste management strategies.

The Government has given waste collection authorities a duty to prepare and issue waste recycling plans, in section 49 of the Environmental Protection Act.

It is against this background that this Waste Management Paper has been prepared. Its primary purpose is to provide guidance to local authorities on devising and implementing recycling strategies. More specifically, it gives advice to waste collection authorities on how to draw up recycling plans.

Chapter 1 introduces the Paper.

Chapter 2 describes how recycling fits in as a waste management option. It describes why it is becoming an increasingly important option and spells out the basic requirements for any successful recycling scheme. It sketches out the legislative and administrative framework within which recycling schemes operate.

Chapter 3 explains how to draw up a waste recycling plan, including the statutory requirements and key factors to be taken into account. It is supplemented by **Appendix A**, which sets out a model outline of a recycling plan.

Chapter 4 provides an overview of the composition of household waste and assesses what proportion is recyclable. It looks at the different types of collection system and stresses the importance of identifying markets for recovered materials and using purchasing power in favour of recycled products.

Chapter 5 summarises the current state of the recycling industries in the United Kingdom, dealing with each of the main materials in turn. **Appendices B and C** provide supporting information about the recycling industries. This chapter also describes the main forms in which waste can be used to derive energy—further information on energy recovery options is at **Appendix D**.

Chapter 6 outlines the economic and technical barriers to recycling.

The Waste Management Paper contains a list of organisations who can be contacted for further information (**Appendix E**) and a bibliography (**Appendix F**).

CHAPTER 1

Introduction

Government Recycling Policy

1.1 The Government is committed to increasing the level of recycling of household waste. It has set a target to recycle half of the country's recyclable household waste (about 25% of all household waste) by the year 2000.

1.2 The White Paper on the Environment, 'This Common Inheritance', issued in September 1990, states that recycling can make good environmental and economic sense, given the right conditions. It will conserve natural resources; save energy in production and transport; reduce the risk of pollution as well as saving costs in pollution control; reduce the demand for landfill space and produce goods more cheaply.

1.3 The White Paper points out that no more than 5% of the UK's household waste is currently being recycled. Since materials such as glass, cans, plastic bottles, clothing and newspapers can readily be recycled and vegetable and food waste can be composted, there is scope for a much higher level of recycling. In fact, the White Paper estimates that half of the domestic waste stream could readily be recycled. It sets out a range of measures designed to achieve the Government's target.

The Role of Local Authorities

1.4 Local authorities have a key role to play in achieving the Government's recycling target because they have the responsibility for collecting and disposing of household waste. That is why the Government is placing a requirement on Waste Collection Authorities (WCAs) to draw up waste recycling plans, under section 49 of the Environmental Protection Act 1990.

1.5 Although the requirement to prepare recycling plans is placed on WCAs, they will not be able to carry out their recycling role fully unless they work in co-operation with waste disposal authorities (WDAs) and Waste Regulation Authorities (WRAs). The roles of these different authorities in recycling are described in more detail in Chapter 2.

1.6 Up till now, recycling has been an add-on extra for many local authorities. The message of the White Paper, of the Environmental Protection Act and of this Waste Management Paper is that it must now become an integral part of each authority's waste management strategy. As the standards and costs of landfill rise, recycling will become an increasingly important waste management option.

The Purpose of this Waste Management Paper

1.7 Waste collection authorities face the task of drawing up recycling plans. In some cases, they may have limited experience of recycling. This Waste Management Paper is being issued primarily to assist WCAs in their task, by providing them with background information on recycling and suggesting how they might draw up their recycling plans. It offers a model outline of a recycling plan and some advice about how to prepare and implement a recycling strategy.

1.8 The Paper should also be studied by those in WDAs and WRAs who have responsibility for waste management or who have a brief to encourage recycling. Unless these authorities integrate recycling into their waste management policies and practices, the effectiveness of the WCA's recycling strategy will be hampered. As the Paper indicates, there is scope for WDAs and WRAs to take an active stance on recycling.

1.9 Finally, this Paper should be of interest to those who are involved in recycling, whether in the reclamation industries, the community sector or as producers and distributors of products which end up as household waste. Each of these has a part to play in achieving the Government's recycling target.

CHAPTER 2

Putting Recycling in Context

The Waste Management Role of Local Authorities

2.1 Recycling is only one of a range of waste management options. Although the Government expects it to become a more widely-used option, it should not be pursued in isolation, without regard to the other available options for dealing with household waste. It is not the role of this Paper to offer an evaluation of the different waste management options. Waste Management Paper No. 1, which is being revised, already provides such an evaluation. However, authorities considering the place of recycling in their plans should bear in mind the following general points.

2.2 Local authorities need to provide a safe, robust and secure system for dealing with waste. This system must minimise risks to human health as well as damage to the environment. It must also be able to respond to changes in society, in particular to changing waste compositions and changing attitudes to waste. Authorities will almost certainly adopt a combination of options for handling waste, tailored to their requirements and circumstances; the combinations used will vary over time, as well as from authority to authority. Variations will reflect the availability of suitable landfill sites, waste processing capacity, and local market demand for reclaimed materials.

2.3 The principal options available for waste management are listed below. The list represents a broad hierarchy of preferred solutions:—

WASTE MINIMISATION

RE-USE

MATERIALS RECYCLING

ENERGY RECOVERY

LANDFILL INCINERATION

(without energy recovery)

2.4 Although this Paper concentrates on recycling, it is clearly better to avoid or **minimise waste** in the first place, where possible; or to **re-use** a product. Waste minimisation and re-use are largely controlled by industry but some authorities are now thinking of ways in which they can encourage householders to minimise the amount of waste which they generate; or to influence local industry and business.

2.5 Adur District Council, for instance, has distributed a leaflet to householders, suggesting ten ways in which they might minimise waste. The message that waste avoidance or re-use are preferable is an important one for authorities to communicate and to follow in their own practices and policies.

2.6 **Recycling** differs from re-use because it involves a processing step. It is important to recognise this step because recycling is often used to describe only the collection of materials separated from waste. This operation is more correctly defined as reclamation. Recycling can be defined as the collection and separation of materials from waste and subsequent processing to produce marketable products. These marketable products can be in the form of:—

•   materials such as paper and board;

•   finished products partly or totally consisting of recycled materials;

- fuel—solid, liquid or gaseous;

- energy—heat or electricity; and

- compost.

2.7   In the past, the need to provide a robust waste disposal system has directed the waste management industry to rely on the **landfill** and **incineration** options. These are capable of accepting a wide variety of materials and allow the waste management industry to control the operations and resources itself. Where appropriate, materials and energy recovery have been incorporated into the system to generate revenues and reduce costs but the main aim has been simply to dispose of the waste.

2.8   In the move towards recycling, local authorities and the waste management industry need to retain robustness so that the continuing supply of waste can be handled. In addition, authorities must cater for the volatility of markets for recycled materials and recovered energy, by ensuring that they have suitable options to fall back on. It is highly desirable to have the flexibility to be able to switch between waste management methods. This again points to the use of a combination of waste management options rather than over-reliance on one.

Why opt for recycling?

2.9   This section describes the pressures to increase recycling and sets out why it will need to be considered as an important waste management option over the next few years.

2.10   The amount of waste generated in the UK is considerable—in the order of 700 million tonnes/year including agricultural wastes. However, for local authorities, the main focus is on household waste, of which 14–15 million tonnes is collected by door-to-door collections and about 4 million tonnes through 'civic amenity' sites and special collections. It is estimated

that between 80 and 90% of this waste is sent to landfill and 5–10% incinerated or processed to recover refuse derived fuel. The remainder, which consists largely of scrap metal, paper and glass, is collected for recycling.

2.11   There is increasing concern (not only in this country but in the rest of Europe and in North America) about the environmental impact of waste management procedures. The European Community developed a set of environmental objectives incorporated into the first amendment of the Treaty of Rome in July 1987 and a waste management strategy has been evolved from these, which follows the hierarchy described in paragraph 2.3 above. The strategy involves:—

- preventing and reducing waste at source;

- increasing recycling and the re-use of materials and products; and

- disposing of unavoidable wastes safely.

2.12   As part of the programme based on this strategy, the European Commission is preparing a series of Directives which envisage a much higher level of recycling and re-use of products, covering products and materials such as batteries, packaging, tyres and chlorinated solvents. As these Directives come into force, they will have considerable impact on recycling practices in the UK.

2.13   British Government policy will also increase pressures and incentives to recycle. The very existence of a recycling target is significant. As the Environment White Paper indicates, the Government is not only promoting measures to induce local authorities to undertake more recycling (including recycling credits, recycling plans, this Waste Management Paper and the availability of supplementary credit approvals (in England) for recycling investment), it is also:—

- pressing industry to increase its capacity to process and market recycled materials;

- providing funding to assist technological advances and improved design (to make products more recyclable);

- adopting a policy of discrimination in favour of recycled products;

- encouraging industry to reduce unnecessary packaging of consumer goods and to take a measure of responsibility for dealing with packaging waste;

- monitoring pilot recycling schemes and assessing their results, to help shape future policy;

- examining further ways to create incentives for recycling; and

- offering voluntary organisations involved in recycling the chance to bid for grants from the new Environmental Grant Fund. (In Scotland and Wales the Special Grants (Environmental) Programme and the UK 2000 Programme can assist suitable projects. In Wales projects may also qualify for assistance under the urban programme).

2.14 The Environmental Protection Act provides further important pressures to recycle, in the form of the duty of care and higher landfill standards. The Act introduces a strict environmental regime for disposing of waste, one of whose effects is that the standards of landfill management will be raised. For instance, the landfill operator will be responsible for managing the site not only while it is being used but also as long as is necessary after the site is closed. He will also face requirements for higher engineering standards, through the new licensing system. As a result, landfill costs are expected to rise significantly in many areas.

2.15 The cost of disposal to landfill is in any case set to rise in some areas because of the difficulty of locating landfill sites close to major centres of population and the lack of suitable sites. The added costs will include the transport costs involved in the transfer and movement of waste to more distant sites. The result of increasing landfill costs is that recycling becomes more economically attractive, especially when the cost of landfill is reflected in a recycling credit paid by the waste disposal authority to a waste collection authority or other recycler.

2.16 Finally, public pressure and concern provides an incentive to recycle. Environmental pressure groups have raised the public's consciousness about environmental issues and, in turn, the public are applying pressures to central and local government. The introduction of recycling schemes is one way in which an authority can demonstrate its sensitivity to environmental issues.

Basic requirements for recycling

2.17 Recycling consists of four inter-related activities:

- collection and transport of the wastes;

- separation and clean-up;

- processing to obtain marketable materials and products; and

- marketing of the products.

2.18 The success of recycling schemes depends critically on economics. The economics of a scheme depends in turn on the way in which these different stages inter-act. For instance, separating materials early in the waste stream is likely to yield less contaminated materials but it is also likely to be more expensive. The authority will need to discover the extent to which the processor can tolerate contamination, the effect which different degrees of contamination will have on the price obtainable and so on. It will sometimes be worth spending more to produce a more valuable material.

2.19 Therefore, although the local authority is most likely to be involved directly in the collection and separation stages, it will need an understanding of the processing and marketing stages as well.

2.20 In identifying process industries capable of recycling materials, it is important to understand their structure. The recycling operations associated with paper and board, iron and steel, aluminium and glass are established industries where production capacities are generally matched to supplies of wastes and markets for recycled products. Expansion of capacity is often in large steps rather than by a gradual increase, and requires considerable investment over long time-scales. For instance, the planned de-inking plants at Aylesford in Kent and Gartcosh in Lanarkshire each require investment of over £200 million to expand waste paper processing by over 400,000 tonnes/year and £28m is being invested in a new plant at Warrington to provide 50,000 tonnes per annum of aluminium ingot from recycled drinks cans. As a result, the processor must have confidence that the supply of materials can be rapidly expanded through the growth of the collection infrastructure, and that there will be considerable growth of the market for products made from the reclaimed materials, either in the UK or abroad.

2.21 Secure and stable markets are important for the long-term viability of recycling schemes. For recycled materials to compete, they must often be sold at lower prices than virgin materials, although the demand for recycled paper has now led to a premium being paid for some grades of this product. The competition between virgin and recycled products means that prices are controlled by those of virgin materials. If the supply of waste exceeds the demand for recycled products, the waste must be exported, recycled to recover energy, or disposed of to landfill.

2.22 This re-emphasises the need for recycling operations to have the built-in flexibility to use these options, since the generation of waste cannot be switched on and off to suit demand.

2.23 When demand for recycled products exceeds the supply of waste, processors have the option of increasing supply by promoting collection schemes or by importing scrap, or reducing demand by increasing prices.

The Legislative Framework—England

2.24 The main piece of legislation covering recycling is the Environmental Protection Act 1990 (the 'EPA'). Part II of the EPA replaces and extends the provisions of the Control of Pollution Act 1974 relating to waste collection, disposal and regulation. It also introduces some fundamental changes for the collection and disposal of waste.

2.25 WCAs have a duty to arrange for the collection of household and, if requested, commercial waste. Section 49 of the Act puts them under a duty for the first time to produce waste recycling plans (see paragraph 3.2).

2.26 Section 46 gives powers to WCAs to require householders to place waste for collection in specified kinds and numbers of receptacles; separate out waste for recycling and waste not for recycling into separate compartments or receptacles; use receptacles of a certain size and type; place them in a particular place for ease of collection; and certain related powers. Section 47 gives similar (but not the same) powers in respect of commercial and industrial waste. These powers strengthen existing provisions in a way which will assist effective kerbside collection.

2.27 Section 48 exempts the WCA from the general duty to deliver collected waste to the WDA, where it intends that it should be recycled. The WCA must inform the WDA of its plans to which the WDA may object if it already has a contract for the waste to be recycled. Unless the WCA has combined collection and disposal functions, it may also provide plant or equipment to sort or bale waste for recycling. Even if it does have the dual function it may still employ a contractor for that purpose.

2.28 WDAs have the duty to arrange for the disposal of all controlled waste collected in their areas by WCAs. They must also provide places for the public to deposit waste (e.g. 'civic amenity' sites). The EPA introduces a framework for competitive tendering for waste disposal.

2.29 In future, the WDA must carry out its functions through contractors, which may either be local authority waste disposal companies (LAWDCs) or private contractors or joint ventures. This will mean that the true costs of disposing of municipal solid waste cannot be hidden within the authority's general expenditure. This move will make true disposal costs apparent which is, in itself, a vital element of developing a framework within which recycling can thrive.

2.30 Paragraph 19 of Schedule 2 to the 1990 Act requires WDAs, in preparing to enter into waste disposal contracts, to have regard to the desirability of including terms and conditions in the contract designed to maximise the recycling of waste under that contract.

2.31 Section 55 gives powers to WDAs and WCAs for recycling waste.

2.32 Under the EPA, WRAs have the responsibility, under section 50, to prepare waste disposal plans. In drawing up plans, they will be obliged to have regard to the desirability, where reasonably practicable, of giving priority to recycling waste; and to consult the WCA. The WRA will also be obliged to include forecasts of waste arisings and composition in its area.

2.33 The EPA also introduces a system of recycling credits. Where a WDA avoids disposal costs as a result of recycling by a WCA, it must pay the WCA a credit equivalent to its saving. The same applies where the WDA saves the WCA collection costs. Both authorities have a power to pay credits to third-party recyclers. The Government will be issuing separate guidance on recycling credits. The legislation on the payment of recycling credits between authorities does not apply where the same authority carries out waste collection and waste disposal, as in Scotland and Wales.

2.34 The new legislative framework for waste management in the 1990 Act includes other important measures which are significant for recycling. Arguably, the most important consequence of the Act for recycling is that it will raise the standards required for landfill sites, thus increasing the costs of waste disposal. This, in turn, will make recycling a commercially viable option in some cases where it had not been previously. It will impose a duty of care on those dealing with waste and introduce improvements to the waste management licensing system. Regulations under the Control of Pollution (Amendment) Act 1989 will introduce registration for carriers of waste.

The Administrative Framework—England

2.35 From the statutory provisions described above, it will be apparent that the roles of the different authorities are inter-dependent. The most effective recycling strategies will actively involve all three authorities. Authorities should therefore expand and consolidate their links and avoid the danger of duplicating effort. One idea, adopted by an increasing number of authorities, is a county-wide recycling forum, bringing together representatives of the county and districts. The Local Authority Recycling Advisory Committee provides a national network for the exchange of information.

2.36 The links between authorities will include the exchange of information and ideas and the provision of statistical data (of which, more in paragraphs 3.20–3.27). They should also look for opportunities for joint initiatives, to take advantage of economies of scale. Conversely, they will need to ensure that, where they are involved in separate recycling schemes, they operate satisfactorily in parallel. For instance, the inter-action between WCA materials collection schemes and WDA centralised processing schemes will need careful consideration.

2.37 WDAs and WCAs will need to organise arrangements for handling recycling credits and for maximising the benefit from them. Critically, this will involve ensuring that contracts are structured in such a way as to yield disposal or collection cost savings where waste is diverted for recycling. One of the most valuable actions which a WDA can take to promote recycling is to draw up contracts which yield a saving when the amount of waste for disposal reduces, to enable them to save what they pay out in recycling credits.

2.38 In future, recycling needs to be treated as an integral part of overall waste management strategies. This being so, WDAs and WRAs have a vital role to play in ensuring that their plans and activities incorporate recycling, as the provisions of the Environmental Protection Act encourage them to do.

The legislative and administrative framework—Scotland

2.39 The EPA applies to Scotland but with some modifications to reflect Scottish circum-

stances. In particular, WCAs and WDAs are integrated in Scotland (both being functions of district and islands councils). There is thus no requirement for recycling credits to be paid between them, nor do the provisions of section 48 apply to them. Most of the advice in paragraph 2.35–2.38 above will not apply. There is no requirement for LAWDCs to be established in Scotland. However, the following provisions mentioned in paragraph 2.24–2.34 do apply to Scotland:—

- recycling plans (s.49)

- receptacles (s46 and 47)

- recycling powers (s56 for Scotland)

- waste disposal plans (s50)

- recycling credits for third parties (s52)

- duty of care (s34)

- improved waste management licensing system (s35–44).

The legislative and administrative framework—Wales

2.40 In Wales the functions of collection, regulation and disposal of controlled wastes are the responsibility of district councils. The EPA applies to the district councils with exceptions to reflect the unitary nature of their responsibilities. As multi-purpose authorities, district councils are responsible for both the preparation of waste recycling plans under Section 49 and disposal plans under Section 50. District councils will make appropriate arrangements to ensure that these functions are coordinated and that steps are taken to integrate recycling into their waste management policies and practices.

2.41 Coordination between authorities will be undertaken through the regional waste management groups and the scope for joint initiatives will also be considered in this forum. The EPA provides district councils with the opportunity to develop recycling strategies and to undertake recycling schemes through contractors, which may include local authority waste disposal companies.

2.42 The sections of the EPA on the payment of recycling credits only apply to Wales in relation to powers in Section 52(3) and (4) to make payments to third parties.

Co-operative working

2.43 One of the main messages of this Waste Management Paper is that local authorities cannot recycle on their own. They are dependent not only on each other but also on others who participate in the recycling chain.

2.44 The voluntary sector has a long history of involvement in recycling projects and continues to play an essential role. Community groups can help to encourage the general public to collect and sort their recyclables, particularly if the group obtains revenue, either from selling materials or by sharing local authority revenue from 'bring' systems. They offer enthusiasm and motivation, qualities often needed to set up a scheme which endures. They can play a key role in publicity, campaigning and education. They can also adopt recycling sites, keeping them tidy and attractive (see paragraphs 4.21–4.28).

2.45 The reclamation industries are also key players; without them, collection would not become recycling. Industry's role is not limited to buying the materials collected, processing them and turning them into goods to be marketed. There may be scope for a collaborative venture in which a company provides and operates reclamation facilities, finds markets or operates the recycling collection. Authorities should always ask the question: do I need to do this myself or is there someone who could do it better?

CHAPTER 3

Preparing a Recycling Plan

Introduction—The statutory requirements

3.1 This Chapter provides advice on how to prepare a recycling plan, and suggests the different issues which should be considered and the topics which should be covered. Appendix A offers a model outline of a recycling plan.

3.2 Section 49 of the Environmental Protection Act places a duty on Waste Collection Authorities to prepare and issue waste recycling plans. The duty has two main elements: it requires a WCA to carry out certain tasks in relation to the plan; and it requires certain information to be included in the plan.

3.3 The tasks which the WCA is statutorily required to undertake in respect of the household and commercial waste in its area are:—

• **to carry out an investigation** with a view to deciding what arrangements are appropriate for dealing with the waste by separating, baling or otherwise packaging it for the purpose of recycling it;

• **to decide what arrangements are needed** for that purpose in the authority's opinion;

• **to prepare a recycling plan,** which is to be a statement of the arrangements made and proposed to be made by the authority and others for dealing with waste in those ways;

• **to carry out further investigations** from time to time with a view to deciding

what changes to the plan are needed; and

• **to make any modification** of the plan which the authority thinks appropriate because of its further investigations.

3.4 The EPA states that the plan shall include information on the kinds and quantities of controlled wastes which the authority expects to collect and purchase during the period of the plan and the kinds and quantities of controlled wastes which the authority expects to deal with by recycling. The waste recycling plan must explain the arrangements which the authority expects to make with waste disposal contractors (or, in Scotland, with waste disposal authorities and contractors) during the period of the plan, together with plant and equipment which the authority expects to provide for sorting and baling of waste. Finally the plan must include the estimated costs or savings attributable to the methods of recycling detailed in the plan.

3.5 After approval of the draft plan by the Secretary of State for the Environment, for Wales or for Scotland as appropriate, the WCA must publicise the plan in its area and provide copies to the relevant WDA and WRA. Once the plan has been completed, it must make arrangements for the public to have access to it.

3.6 Any authority which is serious about recycling would need a recycling plan or strategy, whether or not there was a statutory requirement. A plan provides a framework against which to judge whether a particular recycling scheme is the best way of promoting recycling in the area.

Key Factors in Preparing a Recycling Plan

3.7 If the preparation of the plan is to be more than a paper exercise, it needs to involve considerably more than allocating someone to write a dissertation on recycling in the area. Otherwise, there is a real danger of producing a document which is quickly forgotten and makes no difference in the real world. To avoid this danger, there are several key factors to address in preparing the plan.

Obtain high level commitment

3.8 The first priority is to get commitment at the highest level, both from councillors and senior council officers, to a clear statement of the principles and objectives. The importance of high level commitment is that it demonstrates the importance that the WCA places on recycling, and emphasises the intention to provide the resources for preparing and implementing the plan. It also helps to obtain cooperation of all departments, operating sites and individuals within the organization and provides a strong influence on external organisations and the community served by the WCA. Without this commitment, the recycling plan runs the risk of only partial success or even total failure.

Appoint a recycling co-ordinator and/or committee

3.9 Once the WCA has committed itself through its councillors and senior officers to a recycling plan, it needs to appoint a recycling coordinator and/or committee if one does not already exist; the committee may comprise elected members, officers or a combination of both. The recycling coordinator or committee must be given the responsibility both for formulating the plan, including setting objectives and targets, and for overseeing its implementation. Input is required from representatives from a range of disciplines including engineering, planning, health and safety, education and communications. Representatives must be selected on the basis of their understanding of and support for the commitment of the WCA to a recycling policy.

Communications and education

3.10 Communications and education are vital to help the WCA staff and the public understand the recycling programme and to encourage their participation. Publicity is essential to capture the attention of the public, using mass media and promotional techniques. Education is necessary to develop and sustain public interest and involvement, and includes both formal and informal presentations. In particular, working in schools is a vital element since schoolchildren are a responsive and influential group. The ultimate goal is to change attitudes and behaviour towards the handling of household waste so that recycling becomes routine. Communications and education must be long-term activities for a recycling programme to succeed. They are discussed more fully in paragraphs 3.53–3.64.

Statistics

3.11 Unless an authority has information about current waste arisings and recycling levels and likely future trends, it lacks the most basic tools for preparing and monitoring its recycling strategy. Statistics are the foundation on which the plan is built. If they are inaccurate or unreliable, the authority will to an extent be taking decisions in the dark. Views about the right direction for future activity will be based on anecdote and conjecture rather than fact.

3.12 Authorities need to set up systems for collecting or obtaining the statistical information they will require for their recycling plan. This will involve close co-operation between WCAs,

WDAs and WRAs. It will also require the involvement of others involved in waste management and recycling in the area. Only when this information is available will WCAs be able to decide confidently what their future strategy should be and be able to monitor its success. Paragraphs 3.20–3.27 provide more guidance on collecting statistics.

Targets

3.13   The Government has set a national target for recycling. It encourages local authorities to do the same. Targets provide a tangible and readily understandable signal to the outside world that an authority is serious about recycling. They allow an authority to monitor its progress towards achieving its objectives and to obtain early warning if the strategy is not working and needs to be adjusted. They are also a useful management tool for motivating those involved.

Consultation

3.14   In the course of preparing its plan, the authority should consult all of the key individuals and organizations in the recycling operation—from collector through to marketer of recycled products—to ensure that a sound plan is prepared. Consultation with existing waste disposal contractors and authorities is essential, for they will have to deal with the non-recyclable residues of the recycling scheme. These residues may be significantly different in nature, requiring different treatment from ordinary waste. Consultation with the WRA is of key importance to ensure co-ordination with the WCA's recycling plan and the WRA's waste disposal plan. The WCA should also consider public consultation of its draft plan. The greater the degree of consultation and wherever possible the pooling of resources between authorities, the greater the prospect of engaging all sections of the community in implementing the plan once it has been issued.

Survey of current waste management practices and recycling activity

3.15   The major steps to be considered in formulating a recycling plan are now described in more detail, with a checklist of topics to be assessed. In practice, the process is not the simple one of starting at stage one and progressing mechanically through each stage in order. It will be more of an iterative process, refining work on each stage as it is affected by work on other stages.

3.16   The first step is to undertake a survey of what is currently happening in the area (the first stage in what the EPA describes as 'the investigation'). The survey might aim to provide the following information:—

- details of the controlled waste being generated in the authority's area, including quantity and composition and details of what the authority collects or purchases;

- information about what is happening to that waste, including the proportion taking various disposal routes (landfill, incineration, recycling, etc.) and the organisations currently providing disposal services;

- methods of collection ie types of receptacle, types of vehicles;

- methods of transfer and disposal;

- number and type of households and businesses;

- current recycling activities (eg bottle banks, 'civic amenity' sites, voluntary group collections, kerbside collections) and organisations involved in those activities; and

• the quantities and composition of waste recovered by current recycling activities.

3.17 WCAs should be able to benefit from information gathered by the WDA in preparing its waste disposal plan under the Control of Pollution Act 1974.

3.18 As well as surveying what is happening currently, the investigation will need to include an assessment of future trends on, e.g. the amount and composition of controlled waste expected to be generated in the area and collected or purchased by the WCA over the period of the plan (as required by section 49 of the EPA).

3.19 This survey will provide part of the information which the authority needs before it can assess what is possible in terms of recycling activity in future. It will need to complete the other stages of plan preparation before filling in the remaining survey details. These include the amount and composition of controlled waste which the authority expects to recycle itself or to have others recycle over the same period; the plant and equipment it expects to provide; and the costs and savings involved in implementing the plan.

Developing a recycling database

3.20 It will be apparent that, in order to complete its investigation for its recycling plan, the WCA will need to develop a recycling database. To achieve this, it will need to work in close co-operation with the WRA, WDA and other WCAs in the vicinity.

3.21 Each of these authorities will in any case need similar information for their own purposes. WCAs and WDAs will need information so that they can calculate payments of collection and disposal recycling credits. WDAs will also need improved information as they make the arrangements for putting their waste disposal

out to contract. WDAs who intend to take an active part in recycling will need data for developing their recycling strategies. And WRAs will need information from WCAs and WDAs to enable them to draw up their waste disposal plans which must take full account of waste reclamation and recycling.

3.22 The basic statistics which the WCA may consider would need to be produced each year are:

• total controlled waste arisings;

• within that figure, the amount collected or purchased by the WCA;

• composition of total waste arising, by material;

• the means of treatment of the waste: e.g. landfill, materials recovery, energy recovery, composting; and

• the amount and proportion of each material used productively, broken down into categories such as: recycling by materials reclamation; recycling by composting; incineration with energy recovery; and refuse-derived fuel.

3.23 For the purposes of assessing progress against the Government's recycling target, authorities will wish to distinguish as far as possible between household waste and other waste, when deriving the statistics.

3.24 The question then arises about how these statistics should be collected and by whom. For instance, there is a need for local authorities to move towards weighing waste, or requiring contractors to do so, where they are not already doing it, so that waste arisings can be more accurately measured. There is also a need for better information about the composition of waste. A current proposal to set up a National Refuse Data Analysis Project at Warren Spring

Laboratory would assist authorities seeking to estimate the composition of their waste.

3.25 Figures for waste arisings might be most appropriately provided by the WDA. Also, certain WCAs will have waste which is directed to reclamation and recovery systems by the WDA. In these cases, the WDA would be best placed to measure the tonnage reclaimed, since it will have the contractual agreement with the operator of the facility and the need, in certain circumstances, to calculate waste disposal recycling credits.

3.26 The WRA will also have a role since it will receive information not only from WCAs and WDAs but also from other sources such as industry and reclamation companies. One of its main roles could be to ensure that the database is as comprehensive as possible and that the approach taken by individual WCAs is as consistent as possible.

3.27 Meetings to tackle these issues should be held as soon as practicable, to develop a database and a system of information exchange which meets the requirements of all parties. The WCAs and WDA have the greatest practical use of the data. The WRA may be in the best position to provide a dispassionate assessment of claims for inclusion in the database and the ways in which information is included.

Locating markets (I)

3.28 From the survey of current practices, it will be possible for the recycling coordinator or coordinating committee to assess materials available for recycling and estimate possible quantities. The next stage is to identify opportunities for recycling and, more specifically, markets for materials that can be recovered.

3.29 Paragraphs 4.54–4.57 explore opportunities which exist at the moment and

discusse at greater length the need to locate markets. Appendix E provides the names and addresses of trade associations and others who will be able to offer advice. At this stage, all that is necessary is to emphasise the need to locate markets for materials which are being collected. It is a cardinal rule of recycling that there must be a market for the materials collected. Therefore, the investigation prior to drawing up the plan must include a survey of market opportunities in general and the identification of potential purchasers for specific materials.

Assessing collection and processing methods

3.30 The co-ordinator or co-ordinating committee will need to consider what options for collection and processing are most appropriate for its area and to achieve its strategy. The decision will be affected by a variety of factors including the nature of the area (e.g. rural or urban), the requirements of prospective purchasers of materials and the type of housing and households in the area.

3.31 The main choices for collection are described in paragraphs 4.19–4.51. The choices are essentially 'bring' systems, and 'collect' systems with or without central processing. These systems can be used singly or in combination, although it is necessary to ensure that the systems do not compete for the same waste materials when used together.

Determining equipment requirements

3.32 The recycling coordinator/committee needs to assess the additional equipment required for recycling, particularly collection, separation and baling systems. This equipment may be provided by the WCA, by a waste disposal contractor, or by the reclaimer or processor, depending on the current waste management arrangements and the possible arrangements with the purchaser of the materials. In Scotland the WDA may provide the equipment. However, in England and Wales, where a WCA is also a WDA, the Environmental Protection Act states that it may own but not operate the equipment for sorting and baling waste.

3.33 Possible equipment requirements are summarised in Table 3.1 below. The amount of central processing equipment will depend on the quality of the reclaimed material that is required, taking account of the costs of improving scrap quality by increased processing of the materials.

3.34 As well as discussion with the appropriate WDA, it is important to discuss the potential for joint recycling schemes with neighbouring WCAs to find out if economies of scale can be achieved, for collection, storage and processing, or bulk transport to a processing plant. Arrangements of this kind also have the advantage that the quantities of marketable materials will be increased, providing the potential for greater negotiating power in dealing with potential purchasers.

3.35 It should be noted that the local authority may not need to provide any of the processing equipment where a 'bring' system is

Table 3.1 Equipment Requirements for Recycling

System	Collection Equipment	Processing Equipment
Bring	Bins, skips, banks Compartmentalized lorries Cranes	Front-loaders/elevators Magnetic separators Compactors Balers
Collect source segregated (without central processing)	Household boxes Compartmentalized lorries	Compactors Balers Materials handling
Collect source segregated (with central processing)	Household boxes or bins Compartmentalized lorries	Materials handling Sorting equipment Cleaning equipment Compactors Balers Granulators
Collect mixed (with central processing)	No additional equipment	Materials handling Sorting equipment Cleaning equipment Compactors Balers Palletizers Granulators Composting equipment
Residual Wastes	No additional equipment	Existing waste disposal facilities or Energy recovery via landfill gas, anaerobic digestion, incineration or refuse-derived fuel

being operated, since the equipment is often provided by the private contractor. The equipment needs vary considerably in complexity, and WCAs will be most experienced in materials handling equipment. Advice on the specialized materials and energy recovery techniques must be sought from the WDAs, equipment manufacturers and consultants. Besides equipment, there must be planning of storage and handling areas within buildings, together with access arrangements.

3.36 The Government has recognised that local authorities will, in some cases, need to undertake extra investment in equipment, by allocating supplementary credit approvals (in England) for recycling investment in 1991–92 and later years.

3.37 In Scotland authorities should include any proposals for recycling projects in their annual capital expenditure plans and they will receive sympathetic consideration.

3.38 In Wales resources for recycling are not separately identified. It is expected that local authorities will wish to give increased priority to recycling schemes within available resources.

Financial and economic assessment

3.39 Authorities will want to conduct financial appraisals for each proposed recycling scheme in order to establish which would be financially viable. If, for example, an authority is considering two possibilities—a 'bring' system and a 'collect' system, with source segregation and central processing—the costs of each scheme should be compared with the cost savings which result from recycling. In particular, there should be a comparison with the costs of the existing waste collection and disposal scheme.

3.40 When comparing the costs of the alternatives, an equivalent basis should be used so that, for example, a scheme to recycle 20% of household waste should be compared with the costs of handling 20% of the existing amount of household waste. It would normally be reasonable to expect that the various options would be costed using a number of different assumptions about the overall levels of recycling. In general, a number of recycling options should be analysed and if more than one is shown to be viable, then the scheme which generates the highest net surplus should normally be chosen, in the absence of any other over-riding considerations.

3.41 An analysis should take account of the costs associated with existing waste collection and disposal schemes. These would normally include the following:—

- the labour costs of collection, delivery and disposal;

- the administrative costs of collection, delivery and disposal;

- the capital costs of equipment;

- the operating costs of equipment: fuel, maintenance, insurance, etc;

- the disposal costs: landfill, incinerator fees, etc; and

- the provision of after-care of landfill sites.

3.42 These should then be compared with the costs of the recycling scheme which include:—

- the labour and administrative costs of collecting and delivering recyclables either to a market or a processing facility;

- any residual disposal costs;

- the labour and administrative costs of processing;

- the capital costs of the recycling equipment, notably vehicles, materials, handling and processing facilities;

- the operating costs of recycling equipment, notably fuel, maintenance, insurance for vehicles and central facilities;

- the administrative costs of the recycling scheme, including the costs of promotional, advisory and materials marketing work.

3.43 The revenues and savings from a recycling scheme include:—

- the revenue from the sale of the materials;

- any revenue from Commercial sponsorship of the scheme.

3.44 By reflecting the costs of the normal collection and disposal service in the first (or 'base') option, the impact of recycling credits and internal savings to the WCA in collection costs should already be contained within the calculation.

3.45 The process of appraisal takes into account both immediate and future costs and benefits, using discounting as the means of comparing costs and benefits which accrue at different times. Discounting requires a discount rate. While central government does not prescribe what discount rate local authorities should adopt, the Government itself might in similar circumstances use a real rate. A time period of at least 10 years should be used and up to 20 years is generally defensible.

3.46 Authorities should not expect schemes to show an immediate profit. Many of the benefits only accrue once the scheme is up-and-running (the exception may be large metropolitan authorities where pressure on landfill sites means that waste disposal costs are already high). For example, current evidence suggests that cost savings from the collection of waste will occur only after a scheme has become established and households are recycling 20–30% of their waste.

3.47 All of these figures will be subject to uncertainty. The costs of collection, separation and processing are dependent on energy costs which fluctuate with supply and demand; selling prices for recycled materials and market prospects change considerably, as described in Chapter 6. An appraisal should take these uncertainties into account by way of a sensitivity analysis that identifies the effect such changes would have on the overall viability of the project, having considered a plausible range of values for each uncertain variable and the effect of the analysis on the ranking of various options. This analysis should be carried out for all major variables. As a result of the sensitivity analysis, authorities should get a better idea of the chance of particular projects being run at a loss. Projects where these chances are low are likely to be preferred.

3.48 A guide which shows practitioners how to conduct an appraisal and a sensitivity analysis is 'Economic Appraisal in Central Government' by H M Treasury (HMSO 1991).

3.49 A financial analysis is concerned only with costs and revenues which accrue to the authority itself. An authority which is evaluating a recycling option should be aware that its policies are likely to give rise to other costs and benefits which fall on the community at large. These should also be assessed.

3.50 The wider costs and benefits of recycling include:—

- savings in the use of energy, reducing pollution, notably global pollution.

- the avoidance of the pollution and loss of amenity which may be caused by waste disposal; and

- any damage to amenity from the recycling facility.

3.51 When considering the results of the financial appraisal, authorities should keep in mind these wider costs and benefits. They are unlikely to able to attempt a full cost-benefit analysis (by putting values on them). But it will be helpful to list the wider environmental impacts and to consider whether they tip the balance of the decision so that the authority may be willing to bear a financial loss in order to generate wider benefits.

3.52 WCAs should note that section 49(3) of the EPA requires them to include in their recycling plans information on the estimated costs and savings which will flow from implementation of the plan.

Publicity, consultation and education

3.53 As explained in paragraphs 3.7–3.14, publicity, consultation and education are very important in determining the success of a recycling scheme. The recycling coordinator and committee must promote the plan, first of all to the staff who are going to be trained to implement the collection and process phases and, secondly, to the general public whose cooperation is required to separate and deposit their waste materials to match the requirements of the collection and separation systems. Consultation helps to engage support, since people feel that they are being involved in the process rather than being victims of it.

3.54 The publicity and education programme needs to include the authority's staff and contractors, to help them understand what is expected of them and how their jobs will be affected. In encouraging active involvement, it is important that the commitment is visible, through news-sheets and briefings to staff. Training should be provided to instil new practices and attitudes, and can itself show the authority's commitment. Performance targets and incentives can also be highly effective in motivating staff. To sustain commitment, staff must be able to see tangible results from their efforts. Objectives must be backed by a system of monitoring the recycling scheme combined with senior management response which recognises good performance.

3.55 In communicating with the general public in the community, the WCA must follow a number of general principles. Public announcements must be backed by actions. All communications must be honest. Targets and claims should be presented with restraint. Optimistic or pessimistic claims will lead to the loss of credibility, and opposition from groups with vested interests. Achievements and problems need to be presented in a balanced realistic way.

3.56 Publicity must be considered from the early stages of the planning process. The plan should have short and long range goals and a strategy for implementing them. The focus should be on the role of the community and all its sectors. The publicity plan should ensure that individuals are exposed to the messages about the scheme on a regular, systematic basis. The recycling coordinating committee should ideally have representatives with expertise in publicity.

3.57 The publicity and education plan should be prepared in the following sequence:

- develop the message;

- identify the audience; and

- select publicity and education methods.

3.58 The message to be given to the public includes basic information on the reasons for the recycling programme, the way it will operate and what it will collect, and the material quality requirements and the associated implications for separating materials. Organisations such as the Friends of the Earth UK2000 unit, UK2000 Scotland and Waste Watch offer advice on how to present this information. An important part of the message to the general public is that they can help recycling schemes if they promote the market through the purchase of recycled goods. The education programme should aim to influence purchasers to reconsider the quality standards that they require. For example off-white paper products and coloured glass containers may well be perfectly acceptable for many applications.

3.59 The audience can be identified by dividing the community into sectors; it is important to focus on those which are important for the long term success of recycling in the area. Typically, these categories will be distinguished by the areas where people live, their type of residence, their age and their membership of organisations. Target audiences will include schoolchildren, retailers, employees in business sectors, members of business and professional organisations, voluntary and church groups, and recreational clubs.

3.60 Selection of publicity and education methods presents the most difficult choices, because of the need to balance the effectiveness of the method with the costs. Sponsorship should be explored as one way to subsidise publicity costs. Publicity can be mainly through print and broadcasting. Advertisements and articles in newspapers, newsletters, circulars and magazines are useful to get over the printed message. Press releases and articles provide a cheap method of publicity. News items, features and interviews presented on local radio and regional television can be very effective in publicising the start of schemes.

3.61 At the same time as general publicity announcing the recycling scheme, it is necessary to instruct residents about the ways that they should separate waste and arrange for its collection. Circulars to households provide instructions on an individual basis. These should be reinforced by newspaper adverts and posters, and by presentations to education, civic and youth groups. The publicity and education plans must take account of the need for reinforcement of good practices that are required; this is important after the scheme has been running for several months, when householders may otherwise lapse into their previous practices.

3.62 Publicity and education about recycling should not simply be the province of the WCA. Education authorities can play a crucial role; perhaps the most important audience to reach is schoolchildren. Schoolchildren are a key group both because they are responsive and because they are influential—they influence other family members and represent the adult consumers and waste-creators of tomorrow. If they can be encouraged to adopt practices which avoid waste and facilitate recycling, the long-term future will be considerably more promising.

3.63 WDAs who are also education authorities will wish to explore the possibilities. The new national curriculum provides an excellent opportunity to introduce waste minimisation and recycling issues into schools, since environmental issues lend themselves to the cross-curricular activities which are being encouraged in the national curriculum.

3.64 Although the national curriculum applies only to England and Wales, in Scotland curricular guidelines will take account of environmental issues.

Implementation

3.65 The preparation of the recycling plan will of course involve making arrangements for its implementation.

3.66 Once the steps already described have been taken, the recycling coordinator or committee must then establish practical **objectives** and quantitative **targets**. Quantitative targets may include:

- recovery of recyclables as a percentage of total waste;

- reduction in waste for disposal;

- yield of recyclables per household;

- participation rate for collection schemes;

- materials and energy revenues; and

- total and net costs per tonne of recyclables collected.

3.67 Setting a **timetable** for the scheme must take account of the time required to obtain funding and building facilities for capital projects, eg a central processing plant. The timetable must also take account of the requirements for planning approval which can delay waste handling projects. The objectives and targets must be presented to and approved by the leading representatives and officials of the WCA, and their level of commitment must be maintained. The plan will require revision and refinement over time but the broad objectives should remain unchanged.

3.68 Implementation of the recycling scheme relies on good **preparation** to ensure that equipment and staff are in place, that processes and markets are ready to accept materials, and that the publicity campaign has informed households of their role in separating and presenting waste for collection. The recycling coordinator has the key task of supervising the implementation stage, in particular to check the collection system and to instruct householders on materials separation.

3.69 Once the scheme has started, it is essential to **monitor** participation, recovery rates, effectiveness of the collection method, costs, revenue and public attitudes. Procedures for doing this should be detailed in the recycling plan. The records of collected materials enable an assessment to be made of how well the scheme is meeting recycling targets. The assessment helps to focus on problem areas, such as lack of publicity and instructions, operational problems of collection and processing, and poor enforcement of quality standards of the materials being marketed. Measurement of performance is essential to let the community know how well they are recycling, and can provide an incentive for maintaining or improving performance. However, in making improvements to the scheme, it is essential not to change the procedures to be adopted by the householders continually, since householders will otherwise become de-motivated. It is essential that they accept a recycling practice as a routine.

3.70 Finally, the plan will need to be **up-dated or a new plan** produced from time to time. The authority will need to consider what is the appropriate frequency for up-dating the plan in the light of local circumstances. Some aspects of the plan will clearly need to be brought up-to-date more often than others. For instance, the authority will need to produce new statistics every year. As a general indication, it might be reasonable to modify the plan formally after four years, depending on circumstances.

CHAPTER 4
Implementing a Recycling Strategy

4.1 Once a recycling plan has been prepared, the next stage is to put it into practice. This Chapter provides advice on four key areas of implementing a recycling strategy. It discusses the types and amounts of material which it should be possible to recycle from waste. It looks in more detail at two topics mentioned briefly in Chapter 3: options for collecting and sorting; and locating markets. Finally, it points out the local authority's position of influence as a purchaser.

4.2 This chapter quotes figures for household waste arisings and composition. These figures are based on the best information currently available but further work is necessary to refine and improve their quality. They will in any case not necessarily be representative for an individual area.

What is recyclable—in dustbin waste?

4.3 The amounts of household waste collected by WCAs through the normal refuse collection service vary widely—from 6 kg/week per household to 20 kg/week, depending on the number of occupants, their ages and their lifestyle and purchasing habits. Other factors such as geographical area, method of refuse collection (bin, plastic bag, wheeled bin), proximity to and availability of 'civic amenity' sites or special collection services, domestic heating systems and size of gardens will also have an effect on both the quantities and composition of waste. A weekly average of around 11–14 kg (ie 550–700 kg/year) is typical for households in many mixed residential areas for which measured data are available.

4.4 The pattern of household occupancy has been changing in recent years, with a decline in average household size from 2.71 persons in 1981 to 2.55 in 1987 (Social Trends—Central Statistical Office). This is because there has been an increase in the number of households containing only one or two people—57% in 1989 as opposed to 54% in 1981.

4.5 There are no available data with which to make a quantitative assessment of the effect of these changes but it seems reasonable to infer that, even if the average amount of waste per household has declined, it is unlikely that the total amount of household waste has diminished. Using the 1987 figure for average household size (ie 2.55 persons/household), the present population of the UK represents 22.33 million households. If one assumes that each household generates, on average, 600 kg of waste per year, the total amount of household waste collected by normal refuse collection amounts to about 14.5 million tonnes per year. Local authorities tend to report higher figures because of the co-collection of commercial and industrial wastes.

4.6 The typical composition of collected household waste in the UK, measured by weight, is as follows:

Paper	33% }	
Plastics	7% }	
Textiles	4% }	of this 52%, all but 2%
Miscellaneous	8% }	is combustible
Glass	10% }	none of this 18% is
Metal	8% }	combustible or compostable
Putrescibles	20% }	
(Kitchen Waste)	}	this 30% is nearly all
Fines (dust)	10% }	compostable (inc. paper = 63%).

This analysis is based on the handsorting of samples of mixed household waste, so the weight in each category includes associated dirt and moisture adhering to the materials.

4.7 The extent to which different items of domestic waste can be recycled depends to a large extent on their degree of contamination. Estimates of the amounts of potentially recoverable materials in household waste, based on work at the Government's Warren Spring Laboratory, are shown in Table 4.1. It should be noted that the proportion of clean recyclable material will vary from area to area.

4.8 Clean waste is expressed as a percentage taking account of factors such as contamination by moisture and dirt and the presence of composite and laminate materials which are not recyclable.

4.9 The estimates are based on practical results from separation schemes. The residue of domestic waste, after full recovery of the clean recyclable materials, is nearly 60% of the total.

4.10 These figures do not take account of the scope for using putrescible waste to make com-post, which the White Paper estimates could absorb an additional 10% of dustbin waste. Residues can also be incinerated for energy recovery or processed to recover refuse derived fuel. The use of these technologies requires additional effort and expenditure but they do ensure a significant reduction in the landfill capacity required for the residues, and they are productive uses of waste. They are discussed briefly in paragraphs 5.64–5.86.

What is recyclable—in 'civic amenity' waste?

4.11 It is estimated that a further 4 million tonnes of household waste is disposed of annually at 'civic amenity' sites. The heterogeneous nature of this material makes it difficult to obtain accurate quantitative estimates of composition, and seasonal variations are considerable, garden waste being a considerable proportion of the waste in the summer months.

4.12 Estimates of typical composition figures for 'civic amenity' waste from a city area are presented in Table 4.2, which shows that there

Table 4.1 Potentially Recoverable Materials in Household Waste

Material	Column 1 Household Waste kg/Year	Column 2 Clean waste as % of column 1	Column 3 Maximum amount of clean material available for recycling (kg/year)
Paper	200	60	120
Plastic film	24	60	14
Plastic articles	18	70	13
Textiles	24	50	12
Glass	60	90	54
Ferrous metal	42	80	34
Aluminium film/cans	4	70	3
SUB TOTALS	372	67	250
Other putrescibles etc	228	Remaining quantity of unclean waste for disposal	350
TOTAL	600		600

can be a three-fold increase in the proportion accounted for by garden waste.

4.13 Not only does the proportion increase between winter and summer, but the total amounts taken to 'civic amenity' sites also increase. Other components change less markedly with season, apart from wood and miscellaneous combustibles which increase in the summer, possibly because of do-it-yourself activities. Because of the high amounts of garden waste, composting is clearly an important option for reduction of waste volumes and recovery of a resource. Further details of composting are given in paragraphs 5.64–5.73.

What is recyclable—in commercial and industrial waste?

4.14 Commercial and industrial waste collected by WCAs is generated by a variety of businesses including shops, distribution depots, offices, transport services, catering and public services. The physical characteristics are similar to those of household waste, although the proportions will differ; in addition, the wastes also include discarded office and industrial equipment.

4.15 There is little information from which to estimate the total quantity of waste from these sources accurately. Limited analyses have been made from sources such as offices, airports and supermarkets. These analyses show that office wastes can contain over 80% discarded stationery. Wastes from catering, leisure and similar establishments will of course contain significant amounts of glass, beverage cans, plastics, paper and board packaging, and food waste.

4.16 Greater variations exist in the composition of waste from industrial premises. In most industrial operations, solid wastes are normally sufficiently large to justify separate disposal arrangements. However in small companies, the amounts of process wastes may be very small eg less than 100 kg/week. If these are of a non-hazardous nature, they are usually mixed with general site refuse for disposal. In general, the composition of the wastes closely reflects the types of on-site activity carried out by the waste producers.

Table 4.2. Typical Composition of "civic amenity" Waste Samples.

Material	(Wt %) January	(Wt %) May
Paper and card	5.7	2.2
Plastics	0.7	1.8
Textiles	5.2	2.1
Wood	10.9	4.1
Household (>450mm)*	5.2	8.8
Miscellaneous combustible	4.2	1.4
Miscellaneous non-combustible	23.0	9.1
Glass	1.8	0.7
Garden —small	18.2	54.6
Garden —large (>450m^2)	3.7	9.2
Metal	14.8	2.8
Household waste (bagged)	4.7	4.8
Oil	1.9	0.0

* Large items such as furniture and domestic equipment

4.17   Estimates of the potential annual recoverable arisings from commercial premises are:

- 0.8–0.9 million tonnes of paper and board;

- 0.1–0.2 million tonnes of plastics;

- 0.3–0.4 million tonnes of glass; and

- 0.1–0.15 million tonnes of aluminium.

The major recyclables from industrial premises are paper and board, plastics, and aluminium.

4.18   Commercial and industrial organisations should be encouraged to set up systems and facilities to ensure that staff sort their waste and that storage is available for the reclaimed materials. Otherwise, a large proportion of potentially recoverable materials will be degraded by contamination.

Options for collection and sorting

4.19   The main options for collection have been mentioned in paragraph 3.31. They can be characterised as being 'bring' systems, and 'collect' systems with or without central processing. These choices are not mutually exclusive. In any particular area, a combination of systems may be used. The main advantages and disadvantages of each system are presented in Table 4.3.

4.20   In assessing options, the authorities should examine the scope for partnerships with voluntary groups in collecting and sorting waste; and for partnerships with industry, to provide and operate reclamation facilities.

'Bring' systems

4.21   'Bring' methods are systems in which clean, segregated materials eg. paper, glass, cans and discarded manufactured products eg car bat-

Table 4.3 Summary of Reclamation Systems

SYSTEM	ADVANTAGES	DISADVANTAGES
Bring	Low capital costs. Easy to expand to more categories of materials. Accessible. Proven in UK. Can provide clean, marketable materials. Residue amenable to further processing. Range of container sizes to suit density of sites/population.	Quality and quantity dependent on householder. Rate of recovery limited to 15–20% of total waste. Vulnerable to theft and vandalism. Can become unsightly. Noisy. Acceptable only in certain sites.
Collect source segregated (without central processing)	Convenient for householder. High recovery rates (up to 30% of total waste). Potentially integratable with waste collection, composting and energy recovery.	Higher capital and operating costs. More complex to manage. Slow collection for kerbside sorting. Limited experience in UK.
Collect source segregated (with central processing)	Convenient for householder. High recovery rates (up to 30% of total waste). Potentially integratable with waste collection, composting and energy recovery.	Higher capital and operating costs. More complex to manage. Limited experience in UK.
Collect mixed (with central processing)	No effort for householder. High collection rates. No additional collection costs. Potential technology for recovering energy and compost.	Expensive central processing facilities. Highly contaminated low quality materials recovery, and marketing difficulties.

teries are delivered by the public to collection sites such as bottle and paper banks, 'civic amenity' sites and scrap merchants' yards. They have been widely adopted in the UK for diverting materials from household waste, particularly glass and paper. 'Bring' schemes in this country have been set up by collaboration between local authorities, waste management companies, voluntary groups and materials manufacturers. Glass bottle banks have proved to be popular as is shown by the fact that the amount of glass collected and recycled in the UK in 1990 was 20% of UK consumption of glass. Between June 1990 and June 1991 the number of Save-a-Can banks doubled to 426 can banks nationwide. It has also been demonstrated that the public will similarly support 'bring' schemes for other easily storable items such as textiles and paper.

4.22 'Bring' schemes in this country have been set up by collaboration between local authorities, waste management companies, voluntary groups and material manufacturers. Glass bottle banks have proved to be popular as is shown by the fact that the amount of glass collected and recycled in the UK in 1990 was 20% of UK consumption of glass. Between June 1990 and June 1991 the number of Save-a-Can banks doubled to 426 can banks nationwide. It has also been demonstrated that the public will similarly support 'bring' schemes for other easily storable items such as textiles and paper.

4.23 Collection banks can also be used for materials such as plastic containers and food cans; in these cases, there may be additional difficulties because of low bulk density or contamination of the materials.

4.24 One of the issues to be considered, in introducing 'bring' systems, is the density and distribution of collection banks. Where there is a high concentration of collection points, there is evidence that a reduction of 10–15% in waste can be achieved through such systems.

Significant increases in recovery rates can be obtained by increasing the number of banks. The glass industry is doing this and clearly regards 'bring' systems as a valuable route for increasing glass recycling. Its target is to have 1 bottle bank for every 5000 people in the UK by 1995.

4.25 Siting of collection banks is critically important. The location should be chosen to maximise convenience and accessibility to the householder. The site must be visible to the local community, yet secure from vandalism. Bins and vehicle access to them must be on hard standing (eg car parks) with sufficient space and access both for householders' vehicles and the collection vehicles. Sites should also be chosen to minimize the amount of extra car travel to deliver materials, since otherwise more energy can be consumed than is saved by recycling the waste material. City centre and supermarket car parks are obvious sites for bottle banks, as are local community shopping centres.

4.26 Siting should also take account of the need to provide easily accessed facilities for the non-car owner. In densely populated areas, smaller containers, eg Eurobins, in residential streets may be a more useful option to improve recovery rates. Representatives of companies operating bank schemes can often provide useful advice on siting from their wide-ranging experience.

4.27 Authorities should consider the idea of voluntary groups adopting a site. On the basis of the criteria just described, the three main options for collection bank sites are: car parks; supermarkets; and supervised 'civic amenity' sites. The first two options can lead to untidiness, unless they are properly serviced. Authorities should therefore explore the scope for voluntary groups to manage the sites. In return for a fee calculated on the basis of the rate of collection, a group can ensure that the bins are emptied when full and that the sites are kept clean and tidy.

('Civic amenity' sites are usually manned and are therefore likely to be better managed.) In some cases, supermarkets are willing to accept the responsibility. One way of encouraging greater use of banks is to agree to donate some of the proceeds to charity where this is economically feasible.

4.28 Design of the collection containers is an important factor in attracting householders to recycle and comply with the segregation system. The bins should be standard containers which are immediately recognisable, easy to fill and easy to empty mechanically. They should be aesthetically pleasing—clean and well-maintained. Bottle banks are of three types:

1. Jumbo banks which are essentially waste skips with separate compartments, for example for coloured and uncoloured glass.

2. Modular banks for each specific material or type of material.

3. Eurobins which are essentially side-loading wheeled bins with capacities up to 1,200 litres.

Selection of the type depends on site size and accessibility, intended level of provision and compatibility with the collection system.

'Collect' systems

4.29 'Collect' systems involve householders putting out clean segregated materials for collection separate from normal refuse (eg kerbside collection by local authorities, community action groups etc.) or by sale to itinerant scrap and waste dealers (eg 'totters').

4.30 The first multi-material 'collect' scheme was introduced in an area of Sheffield in 1989. More recently, Milton Keynes, Leeds and Bury have initiated multi-material schemes. Cardiff has inherited a monthly scheme operated previously by a community organisation and 'collect' systems of various kinds operate in North East Fife, Dundee and Falkirk. The principle of these schemes is that the householder separates recyclable materials from the waste and places them in separate containers for collection. The materials are then collected in conjunction with the normal refuse collection round or by a separate vehicle. Four main types of scheme have been tried, varying in what materials they collect, the degree of sorting by the householder, and methods of sorting after collection.

4.31 In the first type of scheme, the householder places all of the recyclable materials in one container, the collector empties the container into the vehicle, and the materials are sorted by physical means at a central depot. In the USA, several separation plants (called Material Recovery Facilities or MRFs) have been built to sort a variety of materials from input streams. The processing equipment varies from a sophisticated physical separation circuit to a simple conveyor belt from which materials are separated by handpicking. Compaction equipment is used to bale the separated materials, so that transport costs are reduced.

4.32 The second system of kerbside collection requires the householder to place recyclable materials in one container for sorting by the collector when the materials are collected. This 'Blue Box' system has been adopted for the Sheffield UK 2000 project to test the viability of the system in the UK. The design of the collection vehicle is important to optimise collection and sorting times, so that the maximum number of households can be serviced and maximum amounts of materials can be recovered. Further work to optimise the 'Blue Box' scheme is continuing in Sheffield, and a subsequent 'Blue Box' scheme in Milton Keynes has changed the design of the collection vehicle to improve collection rates. A large scale trial is also starting in Adur district in West Sussex.

4.33 The third system requires householders to sort recyclables and place each category in separate containers or sections of a container. The collector then empties the contents of each container into separate compartments of the collection vehicle. Variants of this system are being tried in Bury and Leeds. Although it requires motivation by the householder to sort the materials correctly, the collection rate is higher than for the system requiring sorting by the collector. This system being tested in Leeds uses split bins which can be emptied mechanically into dual compartment vehicles. Leeds collect four categories: paper and textiles; cardboard, cans and plastics; organic compostable waste and non-recyclables. Collection of two categories per week allows a simple system of collecting the four categories every two weeks. On return to the depot, the partially segregated materials are sorted, and putrescible materials are composted. A split bin system is also used in Bury.

4.34 The fourth collect system is the back-hauling system, where waste materials are returned up the distribution channel to the materials processor. This system has been proposed in other European countries, and is being tested for plastics collection using milk delivery vehicles in Milton Keynes. In addition, the system is used by some large retailers to deal with their packaging waste. The main barrier to this system is that there must be precautions to ensure no cross-contamination between the products being distributed and the waste being collected.

4.35 As the descriptions above indicate, there is a variety of factors to consider in designing a kerbside collection system. One of the most important factors is the type of collection vehicle to use.

4.36 For box collection schemes, low-loading trucks or trailers are required, with compartments which can be varied in size to match the volumes of separated recyclables that are collected. Specialized vehicles with cabs containing dual controls and low-level access can be obtained, to reduce manning requirements. For many cases, hydraulic tipping or other rapid unloading methods (e.g. roll-on, roll-off containers) are necessary to facilitate speedy unloading. In addition, hydraulic loading can improve collection efficiency. Compaction helps to increase the number of household collections per round before the vehicle must return to the central processing or storage facility though it may cause problems for subsequent sorting. The costs and benefits of these features must be assessed carefully.

4.37 An important factor in vehicle choice is the sorting method to be used. For box collections, the choice is between:

- sorting as the vehicle is loaded;

- sorting at a central materials recovery facility; and

- partial sorting as the vehicle is loaded, with further sorting at a central facility eg. newspapers in one compartment and other recyclables in a second compartment for central processing.

4.38 The first option has low capital costs for central processing or handling facilities but high labour costs for collection. This option has been used in Sheffield and with a three person crew collects from about 1000–1500 households per day. The second and third options require capital investment in central processing facilities and incur operating costs for labour and for operation of the equipment. The amount of equipment and the degree of sophistication in separating and cleaning materials depends on the materials demanded by the market and the quality required.

4.39 Where source separated waste is collected in compartmentalised or multiple bins, it may be possible to modify existing compactor lorries for collection, though experience in some schemes suggests that specially-designed vehicles may be necessary. Recyclables and residual materials can be collected alternately if the householder is provided with two bins. Central processing facilities are required, but again the technical requirements and investment depends on the markets.

4.40 Another factor in designing the scheme is the nature of the collection area. Hills, housing type, narrow streets, and small cul-de-sacs can reduce the rates of collection. Distances between collection and delivery points must also be considered in deciding on vehicle capacity and mechanical specifications. Such considerations will of course be routinely taken into account in specifying vehicles for normal refuse collection. Areas of dense urban housing which may have inadequate storage space for householders will require particular imagination and flexibility when devising the scheme.

4.41 Once collection and sorting methods and equipment needs have been agreed, collection schedules and routes must be worked out in conjunction with existing waste collection schedules. For householders' convenience, collection of recyclables should be on the same day as collection of residual wastes, but if the same crews are used for both collections, different collection days may need to be assigned. Collection crew size and duties must be worked out to provide efficient utilisation of staff and equipment.

4.42 The reaction of UK householders to the kerbside collection has been positive. In Sheffield, there is up to 90% participation with at least 20% by weight of all household waste being recovered for recycling, equivalent to about 50% of clean recyclables. Regular weekly participation rates are about 60%. The level of participation in Dundee's Eurobin pilot scheme for paper and board is about 75% and Falkirk's kerbside scheme has a similar participation rate. Schemes in Germany and Denmark suggest that

at least 25% reduction in waste arisings can be achieved but the more complex the system, the greater the potential for contamination of the 'clean recyclables' by either 'misplaced' material or material that is already heavily soiled.

4.43 For the simple duobin systems in which mixed recyclables and a 'rest fraction' are collected, up to 25% 'dirt residue' in the recyclables has been reported. Collect systems are likely to suffer from greater contamination of the different fractions than bring systems because not all of those participating will have actively volunteered.

Centralised waste treatment of unsegregated waste

4.44 Centralised waste treatment in which unsegregated or only partially segregated materials are processed centrally, can range from large scale mechanical sorting of household refuse mixtures to the dismantling of discarded household appliances to reclaim materials or components for reconditioning or reuse as spare parts.

4.45 Since collect and bring systems lead to greater collection costs, the use of the normal refuse collection system for materials recovery is attractive and there has been considerable effort in many countries to extract materials from mixed waste using centralised mechanical sorting systems. In the UK plants were built at, for instance, Doncaster, Newcastle and Eastbourne. Doncaster was the most ambitious in terms of the range of materials potentially recovered—including glass, paper, ferrous and non-ferrous metal, film plastics and pelletised refuse derived fuel (RDF). In practice, all the plants which continued to operate in the UK eventually concentrated on recovering RDF and ferrous metal. Although there were technical and operational problems with the recovery of other materials, the main limitations were lack of market demand for the grades of the individual materials recovered (due to the level of contamination of the materials) and the high cost of recovery.

4.46 In addition to the recovery of the RDF, interest has grown in treating the putrescible rich fraction in both anaerobic digestion and aerobic (composting) systems to produce compost/soil conditioner. Apart from the recovery of a potentially saleable product, digestion also reduces weight and volume of the putrescible fraction (by up to 40%) and the stabilised material has much reduced potential to form leachate or biogas if it is landfilled.

4.47 Other materials are recoverable from centralised waste treatment facilities, by adding appropriate equipment circuits to the basic waste handling plant. Ferrous metal recovery can be easily achieved, and technically effective methods for recovering aluminium have been developed.

4.48 Recovery of plastic is currently being examined by the Greater Manchester WDA to confirm the technical requirements for effective plastics recovery and re-use. Although the system feed is a reject fraction from a system primarily developed for paper recovery from mixed waste, the studies will be relevant to RDF plants, sorting of mixed recyclables from a collect system and polymer specific separation for any mixed plastic concentrate.

4.49 There are many options for centralised waste treatment and recovery facilities, from energy recovery alone to combined materials recovery, energy recovery and compost systems.

4.50 Where central processing of source segregated or mixed materials is planned, the equipment requirements are more specialized and capital intensive than for bring or collect systems. Since WCAs will not usually have the expertise to build and operate these plants, they will need to discuss the provision of facilities with the WDA.

4.51 Paragraphs 5.74–5.86 and Appendix D discuss energy recovery; in considering this option, the scope for combining it with materials recovery should not be overlooked.

Employment opportunities for disabled people

4.52 Sorting plants have been established in a number of areas which make use of 'picking belts'. Trained operators sort waste by hand usually from separately collected recyclable materials. Schemes operate, for instance, in Milton Keynes, Leeds (dealing with materials from kerbside collection schemes) and Sheffield. The latter is run by the British Plastics Federation and sorts plastics into different polymer types to maximise the recycling potential of mixed plastics collections.

4.53 Some of these plants employ disabled people or people with learning difficulties. Such people can work successfully in such situations and obtain a real benefit or sense of purpose from so doing. Local Department of Social Security offices may be able to assist in obtaining staff for this work who might otherwise be without employment. DSS rules currently allow people, registered with a general practitioner, to undertake therapeutic work and earn up to £35 per week without losing invalidity benefit. From April 1992, a disability working allowance will permit disabled people to do part-time or low paid work without losing benefit. This could help to establish additional valuable employment opportunities in this area of work.

Locating markets (II)

4.54 Recycling will not happen unless there are markets for the recovered materials. Therefore, identifying markets and locating specific firms who will purchase the materials is a vital part of an effective recycling strategy.

4.55 Opportunities for recycling are diverse and it is useful to try to target certain materials where waste supplies and markets are readily available. Some materials are already recycled on a national basis by certain companies and industries, making it easier to set up a local scheme. These schemes are described in Chapter 5, which reviews the recycling of each material. However it is useful to summarise these opportunities, since they are essential to successful implementation of a recycling strategy. The summary, shown in Table 4.4, is based on the situation in early 1991, and will change. It does not claim to be comprehensive.

4.56 Having explored opportunities and identified potential purchasers, the next step is to negotiate an arrangement with a specific purchaser. These negotiations will affect the collection and processing options which the authority chooses, since the purchaser's specifications determine how materials are to be collected and processed. Additional collection and processing costs to obtain higher quality materials must be balanced against the premium that the purchaser will pay for these quality standards.

4.57 Generally, the higher the quality of reclaimed materials, the more market options exist. A wide customer base will help to support recycling schemes in times of materials oversupply and recession. Negotiations with purchasers must cover quality of materials, prices and arrangements for shipment. Prices should take account of the long-term needs of the recycling scheme and the market. Stable, guaranteed markets should not be jeopardised by higher short-term prices. A contract with a recycler on a long term minimum price basis can be advantageous in reducing risks and establishing a stable financial position for the recycling scheme. Waste

exchanges may also help to deal with short term fluctuations in supply and demand.

Using purchasing power in favour of recycled products

4.58 Local authorities who are looking for markets for recycled products should start at home—with their own purchasing policies. As a major purchasing sector, local government has an important role in expanding the market for recycled materials and products.

4.59 Firstly, an authority's own purchases can help to provide a market. Secondly, it can set an example to organisations and individuals in the area to encourage them to buy recycled goods. The local authority associations have published a guide 'Environmental Practice in Local Government', which gives case studies of recycling projects in various areas. The studies show the influence that local authorities have in working with local organisations. By setting specifications to allow the purchasing of recycled products, local authorities can encourage business and the community to revise their standards to include recycled materials. At the moment, specifications are often set to fit the properties of primary materials rather than being based on the minimum standards required for the particular application. In some cases, virgin materials are specified even though current processes mean that recycled materials would be equally acceptable.

4.60 Through its public education programme on recycling, the authority can influence the general public to change their purchasing behaviour. Recycling programmes provide a vehicle for challenging residents to re-assess their habits and change to recycled products.

Table 4.4 Summary of Recycling Opportunities

Product	Organisation	Source
Waste Paper	Paper Recycling Trade Associations Local Waste Paper Merchants	Newspaper and magazines Office paper/trade packaging board
Ferrous Metal	Save-a-Can and others British Steel Tinplate British Scrap Federation AMG Ltd	Can collection skips Magnetic separation of cans White goods vehicles Electrolytic detinning of cans
Aluminium	Aluminium Can Recycling Association British Alcan Aluminium/Alcoa Aluminium Food Container Association British Secondary Metals Association	Can collection banks, recycling centres Collection schemes Foil collection schemes
Glass	British Glass Manufacturers Confederation (British Glass)	Glass banks
Plastics	British Plastics Federation RECOUP	Outlets for plastics (directory of recycling companies) Plastic containers collection
Compost	Bin manufacturers Wye College Henry Doubleday Association	Collection/home composting Composting technology Home composting
Energy	Energy Technology Support Unit Waste Processing Association Combined Heat and Power	Energy from wastes eg refuse derived fuel, anaerobic digestion Refuse derived fuel Combined heat and power
Miscellaneous	British Scrap Federation Chemical Recovery Association The Reclamation Association/Spastics Society/Oxfam British Secondary Metals Association Tools For Self Reliance	Chlorofluorocarbons Oils and solvents Textiles Other non-ferrous metals Tools for third world countries

Information on multi-material collection schemes can be obtained from the following organisations as well as the local authorities who operate such schemes:

Friends of the Earth Projects Unit	Blue Box Scheme
European Recovery and Recycling Association	Blue Box Scheme
Local Authority Recycling Advisory Committee	Split Bin Scheme

CHAPTER 5

The Structure of the Recycling Industry

General

5.1 The recycling industry in the UK is substantial, particularly for ferrous and non-ferrous metal, glass and paper & board. As a proportion of production, aluminium scrap use is high. The individual types of materials are considered below and summarised in Table 5.1.

5.2 The figures presented in this chapter provide simply an overview of the situation for each commodity, for in practice there are many grades and types of materials within these broad classifications, requiring scrap grades to be matched with specifications for new materials and products. In principle, however, the potential flow of recyclables from household waste could have a massive effect on the existing structure and production capacity of several of the UK commodity industries for established materials

Paper

5.3 UK production of paper and board amounted to 4.8 million tonnes in 1990. About 53% of the current fibre requirements of the UK industry are recovered from wastepaper with the balance being supplied in imported (33%) and home produced (14%) pulp. The difference between the 53% quoted here and the 64% in Table 5.1 is due to 'shrinkage' (the loss of non-fibrous material from wastepaper).

5.4 In the same year, UK consumption amounted to 9.3 million tonnes (see Appendix B, Table B1 for details).

5.5 Around 3.1 million tonnes of wastepaper (excluding mill scrap which is recycled in-house) were recovered from all sources in the UK during 1990. Of this amount some 820,000 tonnes represents process waste recovered from printers and product manufacturers. The remaining 2.28 million tonnes comprised post-consumer wastepaper collected from industrial, commercial and household sources.

5.6 An estimated 225,000 tonnes of wastepaper are recovered by local authorities in the UK (about 180 authorities operate paper banks). In addition it is estimated that 120

Table 5.1 UK Materials Statistics 1990

Material	UK Consumption M Tonnes	UK Production M Tonnes	Current Scrap Uses[1] M Tonnes	Current Scrap Use as % of Production (Consumpt.)
Paper & Board	9.27	4.82	3.09	64 (33)
Glass containers	*1.85	1.82	0.37	20 (20)
Plastics	3.51	2.20	0.20	9 (6)
Aluminium	0.52	0.40	0.22	55 (42)
Ferrous (1989)	14.90	19.86	9.00	45 (61)

Note [1]. Includes scrap collected in UK and exported.

* = estimate

thousand tonnes per annum of wastepaper (mainly old newspapers, magazines and brochures) are collected by community groups and bring/collect systems sponsored by wastepaper merchants and mills. Hence the estimated total post-consumer waste from households is 345 thousand tonnes. Industry and commerce contribute a further 1.94 million tonnes.

5.7 The term 'wastepaper' includes wastepaper, waste board, and paper and board laminates with plastics and metals. It can be classified into broad types on the basis of its quality, ie its potentially recoverable fibre and contaminant content. A grading schedule devised by the British Paper and Board Industry Federation and the paper reclamation industry is used to provide a guide to quality. In the UK it is classified into 11 groups on the basis of their type and form—see Appendix C. Some of the groups comprise unmixed paper or board, normally consisting of material recovered from a single source and with little or no contamination present.

5.8 Post-consumer wastes which have been segregated at source, which are normally recovered from industrial and commercial establishments by merchants, are often closely defined. In contrast, much post-consumer household paper and board waste consists of a heterogenous mixture of several types of paper and board materials. Mixtures of waste paper are frequently sorted by merchants to segregate high value paper and board constituents.

5.9 The eleven groups of wastepaper can be divided into two broader categories based on their use:

(i)   as a substitute for primary pulp, for instance, in the production of printings and writings, kraft wrappings and tis-

sues. These groups are referred to as **pulp substitute grades**; and

(ii)   as materials for the production of packaging and board products for which the use of primary pulp is largely uneconomic. These are usually referred to as **'bulk' grades (or packaging grades)** by the wastepaper trade because traditionally they are supplied and used in large quantities.

5.10 Newsprint is used principally as a substitute for mechanical pulp for newsprint production, for some grades of printings and writings and tissues, and for packaging paper board. Small amounts are also utilized in the production of board and wrapping paper.

5.11 Wastepaper is recycled to recover its fibre content. Compared with wood pulp however, a substantial proportion of the purchased weight of wastepaper may be composed of non-fibrous material which varies widely with the wastepaper grade. The term 'shrinkage' is commonly used to define this loss. A typical shrinkage for post-consumer old newspaper and magazines mixtures waste is 15–20% of the purchased waste.

5.12 The recycling process involves pulping the paper, followed by cleaning to remove glues, staples and other contaminants. The product is then de-inked and (in the case of high quality products such as newsprint, tissue and writing papers) bleached.

5.13 The recycling plants require considerable capital investment. Currently, there are plans to invest over £400 million in two de-inking plants in the UK to make newsprint. Much of the focus has been on recycling newspapers, but the future is likely to see higher amounts of office paper being recycled, this material being of much higher quality.

Glass

5.14 The UK container glass industry produced 1.82 million tonnes glass containers in 1990 comprising:

269 thousand tonnes amber glass 14.8%

292 thousand tonnes green glass 16.0%

1.26 million tonnes white (colourless) 69.2%

5.15 In addition about 655 thousand tonnes of flat glass and 85 thousand tonnes domestic glass are also produced annually.

5.16 UK production is supplemented by net imports of unfilled glass containers, imported foods, wines and other products dominated by green and amber glass bottles of wines, beer and spirits.

5.17 Substantially more than 2 million tonnes per year of glass (excluding containers returned for refilling) are used by industrial and commercial consumers and householders for glazing, table and kitchen ware, as well as containers for foods, wines and spirits, soft drinks, pharmaceutical chemicals, beer and cider, milk, toiletries and other products. The types of glass found in waste vary widely in chemical composition and physical properties eg leaded-glass from heat resistant kitchen ware, glass from glazing and colourless, green and brown glass from food and beverage containers. Most glass waste from households is in the form of discarded containers (bottles and jars).

5.18 The total quantity of glass containers potentially available for recovery in the UK is in the order of 1.85 million tonnes of which amber and green glass account for 810 thousand tonnes (44%) and colourless glass 1.04 million tonnes (56%).

5.19 Details of container and flat glass recycled in the UK in 1990, provided by the British Glass Manufacturers Confederation (otherwise known as British Glass) are summarised in Table 5.2. In addition a further 8,500 tonnes of cullet were exported. Since bottle banks were introduced in the UK in the late 1970s, Leeds and Edinburgh have been the leaders in tonnages recovered, totalling 15,000 and 13,000 tonnes respectively up to 1989. A large amount of glass is also re-used in the UK as refillable containers for beverages such as beer, lager and milk bottles. According to British Glass, 614 million returnable containers were in use by the dairy industry and 201 million for other uses in 1988.

5.20 Glass recycling offers a chance not only to recover the material but also to save energy. The energy recovery occurs in the production of the glass from the cullet. The saving is 4.8MJ/kg (after allowing for transport and processing, this reduces to about 3.9 MJ/kg). Since the average car consumes about 5MJ/vehicle mile, energy savings can be rapidly eliminated by special journeys to return bottles.

5.21 The amount of waste glass which can be recycled by glass manufacturers depends on the desired colour of their products and the colour of the waste cullet available. Ideally, waste glass should be segregated into three colours ie white (colourless), amber, and green for subsequent utilisation. In many instances this is done but frequently in the UK waste glass is collected in only two fractions ie white (colourless) and mixed coloured. Moreover in practice nominally colourless glass as collected often contains coloured glass and cannot be used in colourless glass products (see Chapter 6—Barriers to Recycling). In the absence of adequate sorting techniques, whole batches have to be completely rejected or are only fit for green glass production.

5.22 Concern by glass producers at the increased incidence of mixed cullet through cross-contamination and over-filling of bottle-banks, and poor housekeeping practices at collection sites and storage depots have resulted in several new initiatives to improve standards. For instance, an 'Adopt-a-Bank' pilot scheme to encourage greater glass recovery and enhanced colour segregation has recently been concluded by United Glass in the London Borough of Sutton. A similar scheme is currently being planned for Inverclyde District in Strathclyde where it is hoped to reach recovery methods of 30%. The project, which used wheeled Eurobins on 16 sites, involved the separation of glass into green, amber and clear glass by using different bins for each colour. In this scheme voluntary groups adopted banks, kept sites tidy and generally promoted use of the banks in return for payments related to the amount of glass collected. Collection of the glass was carried out by the local authority using a purpose-built vehicle equipped with side-lifters and two internal compartments. This scheme was deemed a success by United Glass and it has since been adopted by at least five other local authorities.

5.23 Apart from the technical barriers of recycling glass, market barriers exist because of insufficient demand for coloured glass containers in the UK. This issue is described in more detail in paragraph 6.25.

Metal—Ferrous

5.24 Details of ferrous metal production in the UK are provided in Appendix B, Table B5, together with information about recycling levels. The levels of ferrous metal scrap that are used in iron and steel-making processes vary considerably with the refining method and the final product; these are summarised in Appendix B, Table B4.

5.25 Ferrous metal discarded from households and industrial and commercial refuse occurs as components of manufactured goods, as a substrate in materials which have been electroplated with tin, copper, nickel or otherwise coated with zinc, plastics and paints. It comprises cast irons, carbon and mild steel and alloy steels, eg stainless steel and tool steels, mild steels with alloy properties eg microalloyed, high strength low-alloy steels, as well as tinplate, galvanised zinc products and other coated materials mentioned above.

5.26 Warren Spring Laboratory's analyses of hand sorted household waste indicate that about 6–8% (by weight) of the waste is magnetic,

Table 5.2 Container and Flat Glass Recycled to Container Production in the UK During 1990 (tonnes)

| | Tonnage of Cullet | | | | |
	Clear	Green	Amber	Mixed	Total
1. General Public and Commercial (via Councils and Contractors) - Bottle Banks	65,300 (27.2%)	87,300 (36.3%)	18,400 (7.7%)	69,300 (28.8%)	240,300 (100%)
2. Industrial Sources (Bottler and Packers)	15,900	4,900	3,400	10,600	34,800
3. Flat Glass	74,900	22,100	–	–	97,000
Total	**156,100**	**114,300**	**21,800**	**79,900**	**372,100**

primarily ferrous metal. Estimated quantities of ferrous metal from household waste are:

*   cans:—
    0.48–0.66 mtpa

*   white goods etc:—
    0.25 mtpa

*   (to 'civic amenity' sites)

mtpa = millions of tonnes per annum

5.27 For comparison, of the 8–9 million tonnes per annum of ferrous scrap from all sources (including steelworks own scrap), used in the UK 3–4 million tonnes/year are exported and 5–6 million tonnes are used for making iron and steel.

5.28 Articles, goods and appliances collected by 'totters' are roughly processed—usually by hand—to remove unwanted materials. The remaining material is then sold to processors who recover the ferrous content, along with that from similar items and motor vehicles delivered directly to them, by dismantling, fragmentising (shredding), magnetic separation and other methods.

5.29 Where waste is carried by conveyor belts during transfer or treatment operations before being landfilled, it may be technically viable to recover the ferrous fraction by installing electromagnetic separation. A number of authorities already use this technique. Although less than 300 thousand tonnes per year of untreated household waste is subject to magnetic separation, it is an efficient and simple method of reclamation.

5.30 Where magnetic separation from untreated waste is practised, the magnetic fraction (which may contain upward of 75% ferrous cans) is sold to AMG Ltd who remove the tin from the cans and produce a ferrous scrap product by a hydrometallurgical detinning operation. Tin recovered from the process is sold to UK smelters in the form of tin sponge (82% Sn). The current detinning plant at Hartlepool has a capacity of 18,000 tonnes per year magnetically separated refuse and produces 14,000 tonnes per year of high quality steel product containing 0.03–0.05% tin and 750 tonnes per year low grade ferrous scrap, as well as tin (around 50 tonnes/year) and other non-ferrous metal (mainly aluminium). AMG Ltd is currently building a second plant (capacity 20,000 tonnes per year) to service the South of England and Wales with another under consideration to service central Scotland, although no decision has yet been taken.

5.31 British Steel are keen to help examine the possibilities of incorporating electromagnetics in either existing or future waste treatment facilities. They have developed a computer programme to assess the costs of installing and operating magnetic separation in any specific situation and are encouraging local authorities to examine the possibilities jointly with them and AMG Ltd.

5.32 Where refuse is incinerated, ferrous metal (including cans) can be recovered magnetically from the incinerator ash. The tin content of steel cans is relatively high and since tin diffuses into the steel during incineration it cannot be removed by hydrometallurgical detinning. Nevertheless incinerated ferrous material forms a perfectly acceptable scrap feedstock.

5.33 Research has enabled British Steel to guarantee, via merchants, a market for at least 100,000 tonnes/year of such incinerated ferrous scrap, for direct use in steel and iron making operations.

5.34 Discarded household appliances are collected by retailers, by local authority collections of bulky refuse, and via 'civic amenity' sites. They are then usually sold directly to scrap merchants who fragmentize them in hammer mills together with other bulky ferrous scrap, eg, motor vehicles, and their ferrous metal content is recovered using air classification and magnetic separation.

5.35 The 'Save-a-Can' scheme, which accepts both aluminium and ferrous cans has been expanded by British Steel since it took over the scheme in 1990. The number of can banks has risen to over 400, aiming to reach 1000 by 1994. British Steel seek to involve third party sponsors where possible; there should not be a significant financial commitment for the local authority. Save-a-Can provide the can banks and and pay for transportation and other service charges. The authority is asked to provide a local depot, typically consisting of a 35 metres square hard standing area with a low retaining wall.

Metals—Non-Ferrous

5.36 Non-ferrous metals are widely distributed in household waste as metal articles or as major constituents of many commonly used alloys. The non-ferrous metals most commonly found in household waste, and typical items composed of non-ferrous metals, are shown in Appendix B, Table B6.

5.37 The irregular distribution of non-ferrous metals and difficulties of sampling prevent an accurate assessment of the total amounts of each metal discarded in refuse. Estimates based on published data for non-ferrous metal consumption and partial analysis of refuse suggest that the total quantity discarded annually in household waste could be in the range 200,000–300,000 tonnes. An estimate of the amounts present indicate that aluminium-based alloys could account for over 60% (150,000 tonnes/year) of all non-ferrous metal discarded, copper (including the copper content of alloys) 15–20%, zinc (including the zinc

content of brass and die cast alloys) 5–10%, tin (including tinplate, solder and other alloys) 2–4 % and lead in all forms about 1–3%.

Aluminium

5.38 Almost all of the aluminium used in manufactured products is alloyed with one or more other metals to enhance specific physical and metallurgical properties. These alloys are classified commercially into 'wrought' and 'casting' alloys depending on the process which is employed for their fabrication into saleable products. Wrought alloys are fabricated essentially by rolling and extrusion and casting alloys by traditional foundry operations. Table B7 in Appendix B summarises the production and use of aluminium scrap in the UK.

5.39 Total aluminium production in the UK in 1990 amounted to 404 thousand tonnes. UK consumption in 1990 was estimated at 522 thousand tonnes. Scrap collected was estimated to be 220–230 thousand tonnes/year, of which 81 thousand tonnes was recycled by the UK secondary recycling industry and 94 thousand tonnes was exported.

5.40 An estimate of the total quantity of aluminium cans available for recovery in the UK from domestic and other post-consumer sources is around 40,000 tonnes/year—about 27% of the total aluminium used in household and related applications. The other main uses of aluminium are in articles such as hollow-ware, baking trays and cases, kitchen foil and door furniture.

5.41 Recycling of aluminium is valuable for conserving energy as well as materials. Primary aluminium production is energy intensive, and the displacement of primary aluminium by secondary ingot is estimated to lead to an energy saving of 149 Mega-Joules/kilogramme (MJ/Kg). Taking account of the energy to collect cans from bottle-banks and initial processing, the recovery of aluminium leads to an overall energy saving of 121 MJ/Kg.

5.42 The best use for aluminium cans, therefore, is for recycling as new drinks cans.

5.43 Aluminium can collecting is supported by the aluminium industry through the Aluminium Can Recycling Association (ACRA). ACRA aims to achieve 50% recycling by 1995. It has established an organisation and infrastructure to expand the collection of aluminium cans on a nationwide basis. Aluminium cans are taken by collectors to recovery centres (mainly scrap merchants) where they are paid around 50p per kilo—about 1p per can. By September 1990 some 310 recovery centres had registered under the ACRA scheme. ACRA also collaborates closely with industry, commerce and local authorities on can recovery schemes to raise funds for nominated charities and community projects; and is active in the 'Recycling City' projects in Sheffield, Cardiff and Dundee. It has developed active programmes with more than 20 local authorities and a further 30 are under discussion.

5.44 In addition to ACRA sponsored projects, several ventures have been initiated by members of the aluminium industry to recover aluminium cans and foil from household waste. Alcoa has opened a purpose-built aluminium can manufacturing centre at the Swansea Enterprise Park. This plant has a capacity of 100 tonnes of cans per week and has been designed to handle small quantities of aluminium cans purchased from the public as well as bulk quantities from trade and merchant sources.

5.45 British Alcan Aluminium plc, through its subsidiary Alcan Aluminium Can Recycling have established collection schemes in 8 major towns and cities, ie Birmingham, Leeds, Nottingham, Portsmouth, Richmond (Surrey), Sheffield, Southampton and Warrington. Cans

recovered in these schemes are returned to Alcan's existing factory at Warrington, Cheshire for recycling into new can sheet ingots. In 1991 a new plant will be commissioned at Warrington with a potential capacity to recycle 2.5 billion aluminium cans per year, producing 50,000 tonnes of new aluminium ingot which will be used for making new drinks cans. Recovery practices encouraged by the company include collection by schools and charities, other collection methods such as mobile can banks, fixed sites and Can Banks.

5.46 Recovery of foil (principally food containers) is currently being investigated in a pilot project sponsored by British Alcan Consumer Products Ltd and members of the Aluminium Foil Container Manufacturers Association. This project aims to encourage the collection of clean foil by charities and other local community groups for delivery to designated merchants who will pay for it. Alcan Can Recycling is also offering to pay for all foil delivered to any of its public sites in the Birmingham project areas. Data collected during this pilot project are expected to form a basis for a national campaign to recover aluminium foil. The Guide Dogs for the Blind Association also collect aluminium foil.

Other Non-ferrous Metals

5.47 Although a total of around 75,000–80,000 tonnes per year of non-ferrous metals other than aluminium are discarded in wastes, very little is recovered by 'bring' or 'collect' systems. Their recovery is constrained essentially by their mode of occurrence, the need to dismantle products to remove the non-ferrous components and subassemblies present in small amounts, and the low intrinsic value of the recoverable metal in individual articles or products.

5.48 Non-ferrous metals can be recovered as residue when refuse is processed mechanically to sort and recover other products for re-use or to produce refuse derived fuel. During these operations, refuse is segregated into non-metallic and metallic rich fractions. The metallic fractions can subsequently be treated further to separate magnetic from non-magnetic materials. The latter can be further processed to remove aluminium to leave a final fraction containing other non-ferrous metals.

Plastics

5.49   UK plastics production in 1990 was 2.2 million tonnes. Consumption of plastics was 3.51 million tonnes of which 2.49 million tonnes was of the six main commodity polymers (shown in Table 5.3 below). Plastics packaging constitutes over 1.2 million tonnes, most of which has a lifetime of less than one year, representing a large proportion of total plastics waste. Other major uses of plastics in 1990 included building (26% of consumption), automotive products (11% of consumption), engineering components (8% of consumption), and consumer goods (6% of consumption).

5.50   Typical uses of these polymers include:

•   polyethylene—film applications (e.g. carrier bags, shrink wrap) food containers;

•   polypropylene—containers, household ware;

•   polyvinyl chloride—beverage bottles, film;

•   polystyrene—food (eg yoghurt) containers, insulated cups and trays;

•   polyethylene terephthalate—beverage bottles particularly for carbonated drinks.

5.51   On the basis that soiled plastics waste amounts to about 7% by weight of collected household waste, about 0.98 million tonnes is collected each year. In addition, plastics enter the municipal waste stream as packaging and plastic products from commercial sources, and as components of white goods and furniture deposited at 'civic amenity' sites.

5.52   Plastics recycling in the UK is a relatively new activity although manufacturing waste has always been re-processed. The amount of plastics recycled in the UK in 1990 was about 200,000 tonnes of which at least 70,000 tonnes was post-use waste.

5.53   Energy savings can be derived from recycling plastics because the energy intrinsic in

Table 5.3 UK Consumption of Bulk Polymers in the UK - 1990

Polymer	Total Consumption (1,000s tonnes)	Packaging Consumption (1,000s tonnes)
LDPE & LLDPE	700	543
HDPE	380	276
PP	510	170
PVC	615	85
PS	205	88
PET/PBT	84	60

(Key: LDPE = Low density polyethylene; LLDPE = Linear low density polyethylene;
HDPE = High density polyethylene; PP = Polypropylene; PVC = Polyvinyl chloride; PS = Polystyrene;
PET/PBT = Polyethylene and Polybutylene terephthalate).

the material is recovered, and the energy which would otherwise be needed for processing is saved. For LDPE, the energy for production has been calculated to be 100–110 MJ/kg, but the fuel value of the material is 46 MJ/kg. Other polymers such as PET have total energies of production of 170–220 MJ/kg. Energy recovery from recycling depends on efficient collection and processing; typically the energy consumption for recycling can be 10–20 MJ/kg. Energy savings can be over 100 MJ/kg for polymers such as PET.

5.54   Plastics recycling operations concentrate mostly on processing clean industrial scrap of one polymer type, since capital investment and processing costs are lower than for operations dealing with contaminated post-consumer waste. Recycling of post-consumer waste currently focuses mainly on plastics from single types of application such as plastic film from shopping bags, pallet wrap and agriculture, plastic cups, and car battery cases. However, Glasgow District Council, in collaboration with BP Chemicals, has recently installed 60 mixed plastics waste collection centres for all types of plastic and by the end of 1991 will have increased this number to 100. BP Chemicals' primary interest is in the HDPE fraction, with the other polymer types being sold on to other users.

5.55   Collection schemes for plastics are in the early stages of development, with schemes being introduced for bottle banks and kerbside collection by RECOUP, an association of plastics producers and users. In Sheffield, kerbside collection can recover about 70% of plastic containers. Recovery rates from bottle banks are more difficult to measure but are likely to be considerably lower. These source segregation schemes enable cleaner plastics to be obtained than for central treatment systems, but the collection costs are higher. Collection costs for plastics are high because of their low density and poor crushability. Compacting has been tried, but increases the amount of time required for collection.

5.56   Once collected, plastics must be separated from other materials and contaminants by sorting and washing methods. Currently, the British Plastics Federation and the Department of Trade and Industry are funding a project at Radcliffe, Greater Manchester to examine methods to separate plastics from mixed household waste. After reclamation of the plastics fraction, then ideally plastic types must be separated, since mixed polymers are difficult to process. Identification and sorting systems are under development. The separated plastics are then granulated, compounded and extruded or moulded to form intermediate and finished products.

5.57   Mixed polymer processing can be achieved in a limited number of cases, using specialised equipment and polymer compatibilisers. There is one UK operation, Superwood, which converts mixed plastics into wood substitutes in the form of boards or profiles. Technically mixed polymers may be separated by solvent extraction but there are no commercially developed systems doing this. Chemical methods for reversing the formation of polymers have been attempted for polyurethanes and polyethylene terephthalate (PET). Although technically feasible, capital and operational costs have prevented commercial use of these methods.

5.58   The plastics industry is now embarked on an extensive recycling programme. Public pressure, proposed European legislation on packaging (including plastics waste) together with Government encouragement are additional factors leading plastics producers and users to recycle plastics more intensively. Further schemes to collect, separate and process plastics waste are very likely in the next one or two years. Manufacturers are likely, in particular, to pay more attention to the design of plastic products to aid recycling.

Textiles

5.59   About 0.5–0.75 million tonnes/year of textile waste is discarded in household waste. This is supplemented by additional large tonnages discarded in household furniture at 'civic amenity' sites and as unsold items from secondhand clothing shops. It has been estimated by the Reclamation Association that less than 25% of the available material is reclaimed. Although textile recycling has been a long established industrial activity in the UK, demand for recycled fabric and fibre waste has fallen over the last 25–30 years and many textile reclaimers have ceased trading.

5.60   Post-consumer textile waste comprises a complex mixture of natural and synthetic fibres. As a result, the identification, sorting and grading of textile waste for various end uses are labour intensive operations and textile wastes are frequently exported to countries where sorting costs are lower than those in the UK. In 1990 14,290 tonnes of used clothing and other used textiles, 6,576 tonnes of sorted and 4,590 tonnes of mixed rags were exported. There is an expanding demand in developing countries.

5.61   Apart from the market which exists for good quality second hand clothing, the principal markets for old textiles are the flocking and wiper cloth industries. The flocking industry shreds textiles waste for use as fillings in furnishings. This market could expand following the banning of the use of polyurethane foam fillings furniture due to the danger of toxic combustion in products.

5.62   Textile waste for use as wiping cloths must be processed by hand to remove abrasive attachments such as buttons, clips and other fasteners before being cut to size. In recent years, textiles wiping cloths have lost market share to paper and non-woven products. The Wiping Cloth Manufacturers Association have formed a Wiping Rag Advancement Group (WRAG) to promote the use of recycled material as a wiping medium.

5.63   Advice on this sector can be obtained from the Reclamation Association and from charities such as the Spastics Society (which has a national collection scheme) and Oxfam (whose shops accept 'rags' as well as used clothing).

Compostables

5.64   Within household waste, compostables are the materials which are made up of biological matter. This fraction constitutes about 30% of household waste and including paper amounts to some 60%. It is amenable to both composting and anaerobic digestion.

5.65   Composting is a process in which micro-organisms break down certain biological materials in the presence of oxygen, at elevated temperatures (>500C) to produce water, carbon dioxide, compost and heat. The microbial activity is sufficient to self heat the compostable material. This thermophilic composting stage is usually followed by a final stage at ambient temperatures when the compost cures or matures for several weeks. This stage is important for destroying phytotoxins—toxic to plants—which are formed at the first stage of composting. It differs from breakdown of biological compounds by anaerobic digestion where, in the absence of oxygen, methane is formed.

5.66   Techniques for composting include static piles, bins, towers and rotating drums. For effective processing there are three main materials requirements. First, the material must have suitable particle size (<50mm) to maximise degradation and yet maintain adequate free space. Secondly, the initial moisture content must be in the range of 55 to 65% by weight. Finally, the carbon-nitrogen ratio must be controlled to allow rapid composting without excessive nitrogen loss as ammonia.

5.67 Most organic materials are susceptible to composting, particularly food scraps, animal wastes and soft plant tissues. However, bone, wood and industrially altered materials such as some paper types, leather and plant stems are less compostable. In household wastes, there will always be materials which are not compostable and hence will lower the value of the final product. Consequently there are four possible treatment options:

i)   composting whole waste;

ii)   composting a mechanically concentrated organic fraction;

iii)   composting   separately   collected materials; and

iv)   encouraging composting by individual waste producers (eg householders, food manufacturers).

5.68 The approach adopted depends on the intended application which can be divided into four main groups in order of increasing value:

i)   composting for waste reduction and stabilisation;

ii)   composting for low grade uses which tolerate contaminants (such as landfill cover);

iii)   composting for large scale uses with a limited   tolerance   of   contaminants (amenity, forestry, agriculture); and

iv)   composting to produce premium grade composts suitable for all uses including horticultural use and the domestic market.

5.69 Experience of composting whole refuse indicates that the untreated product is not marketable. Consequently, mechanically sorted household waste has been used as material for composting. In the UK, sorting plants have been built to recover refuse derived fuel; compostable fractions have usually been landfilled but have the potential to form higher quality feedstocks. Current UK trials will test the feasibility of this approach. Motherwell Bridge Envirotec are currently engaged in discussions with a number of authorities about applying their DANO system to composting municipal wastes. In Northern Europe, operators have had many years of using this approach, mainly to prepare soil conditioners; high heavy metal contamination of the feedstock has prevented more valuable uses being made of the products. To overcome this problem, there has been a strong move to segregate materials at source to form a less contaminated feedstock for composting. However this approach entails extra effort from householders and higher collection costs, and still requires mechanical separation plant to remove contrary materials. As a result, this approach is most feasible when compostables are collected in an integrated system for collecting other recyclables.

5.70 As indicated in Chapter 4, 'civic amenity' sites already receive large quantities of household garden waste, particularly in the summer months, and much of this is compostable. By managing the outputs to the site, for example by providing separate garden waste skips, appropriate feedstocks for composting can be obtained. This operation is already practised in a number of European countries and detailed investigations have been carried out in the UK.

5.71 Some WDAs have already undertaken work composting suitable green wastes and selected commercial wastes. Sales of this material both to constituent authorities for parks use and to members of the public have been started.

5.72 Home composting has become of interest to local authorities as a way of reducing waste volumes. Some authorities are considering the provision of small plastic composters to householders, and companies have developed specific containers. Such schemes already exist in other countries. This approach is of most value for households with gardens where the composting equipment can be set up, and the product can be readily used. Quality of material is variable although it is not necessary to market the product, if the householder can use the compost himself.

5.73 There is little current practical experience of composting household waste in the UK, and the economics, environmental impact and market opportunities are not well researched. Some information is available on the benefits of the compost in terms of soil structure and fertility. Demonstration schemes are still taking place and will provide a better basis for selecting composting options. Given the proportion of the household waste stream which is compostable, there are potentially great rewards to be gained from successful composting schemes.

Energy from waste

5.74 The focus of this paper is on materials recovery. However, energy recovery also has an important role to play in obtaining value from wastes as UK household waste contains about 9–10 million tonnes of combustible materials.

5.75 Energy recovery can operate hand-in-hand with the recovery of recyclable non-combustible materials, such as the magnetic extraction of steel cans.

5.76 Energy recovery is particularly important as a method of resource recovery from heavily contaminated wastes where it is technically difficult to separate and clean materials, or where the costs of reclaiming materials outweigh the economic value of the materials.

5.77 An important use of energy recovery methods is to treat residues after recovery of materials by source segregation methods. Energy recovery also has a role for dealing with products made up of two or more materials physically or chemically bonded together. Typical examples of these intractable materials include tyres, made of rubber reinforced with steel, and plastic composites and laminates.

5.78 The process of combustion of many of the components of household waste guarantees significant reductions of weights and volumes, and effectively sterilises biological components. Residues from combustion are stable and although they can contain high concentrations of chemicals, eg heavy metals, the small volumes facilitate treatment, stabilization and storage.

5.79 The combustion of wastes and fuels derived from them will obviously lead to the generation of the greenhouse gas, carbon dioxide. However, its global warming potential is significantly less than that of methane. Consequently, unless effective methods are used to collect landfill gas for use as a fuel, combustion is preferable as a way of reducing the contribution of waste disposal to the 'greenhouse effect'.

5.80 In examining the options for energy recovery, it is important to recognise that the residues from source segregation schemes will still contain considerable amounts of combustible material. On the other hand, decisions to invest in combustion plant should take account of the possibility that future higher levels of source separation may alter the calorific and moisture content of waste.

5.81   Numerous methods for the extraction of the fuel content of refuse and other wastes have been examined in recent years and some have been exploited in large commercial operation with varying degrees of success. The principal methods by which energy may be recovered from wastes are:

i)   Incineration of untreated refuse with heat recovery;

ii)   Production of refuse derived fuels (RDFs) in shredded or pellet form following separation of combustible materials by mechanical sorting;

iii)   Pyrolysis/gasification; and

iv)   Anaerobic digestion with methane production.

5.82   A summary of each of these methods is to be found in Appendix D but the role of energy from waste as a waste management option is more fully discussed in Waste Management Paper No 1.

5.83   Interest in energy recovery has been stimulated by the non-fossil fuel obligation (NFFO) within the Electricity Act 1989 (which has to date been implemented in England and Wales only). This obligation enables the Government to require electricity supply companies to obtain a specified amount of energy from nuclear and renewable resources. The first order establishing the initial level of renewable capacity which each company will have available by the end of 1998, was laid before Parliament in September 1990. Projects for generating electricity by incineration of untreated waste and refuse-derived fuel as well as landfill gas were included. This first order is being followed by further invitations for bids.

5.84   The energy recovery options have several advantages and disadvantages which must be taken into account in selecting a system which meets the needs of the WDA. Currently the most common energy recovery methods in the UK are incineration and landfill gas recovery. The main advantages and disadvantages of the options are listed in Table 5.4.

5.85   The options are in varying states of development. Incineration, landfill gas generation and refuse derived fuel are technically proven systems, although increasing environmental standards mean that the methods require better engineering to reduce and prevent emissions and discharges. Incineration is subject to strict emission standards and refuse derived fuel combustion is also likely to need to meet demanding standards although these have yet to be fully specified. Landfill gas generation is subject to Government guidelines (Waste Management Paper 27) and a proposed EC Directive, which require high standards for the engineering and operation of landfill sites. Since wastes have a long active life-time in these sites, control and maintenance is required for many years after the site has been filled. The Environmental Protection Act places the responsibility for maintenance on the site operator, both during landfilling and after the site has been filled.

5.86   The other energy recovery techniques—pyrolysis, gasification and anaerobic digestion systems have not fully been developed, although pilot and full scale trials are in progress. Further time is required before the contribution of these methods to energy recovery can be fully evaluated. Their most practical application may be for specific components separated from household wastes, since treatment of the unseparated wastes can lead to contamination of the processes and the products.

Demolition and construction waste

5.87 It is ironic that surveys of recycling often completely ignore demolition and construction waste, since the processing and re-use of demolition and construction waste can lay claim to being the country's largest controlled waste recycling activity in tonnage terms. In descending order of tonnage, the materials themselves will comprise:

- soft wastes, such as soils and sub-soils;

- hard materials, including concrete and bricks (where not reclaimed whole for re-use);

- a miscellany of timber, plasterboard, metals and other detritus mainly from building operations.

5.88 The greatest potential for re-use exists for the harder materials which, wherever possible, should be kept separate during demolition process.

5.89 Hardcore materials should be segregated at 'civic amenity' sites, whenever practical, so that they can be re-used, for example, in landfill

Table 5.4 Status of Energy Recovery Techniques

Technique	Advantages	Disadvantages
Incineration	Maximum reduction of waste volumes, maximum energy recovery, provides heat and energy.	Capital intensive plant, requiring sophisticated abatement methods; needs a large source of waste close to the incinerator.
Refuse Derived Fuel	Recovers about 30% of total household waste as storable transportable fuel, flexible system capable of recovering materials as well as energy, capacity can be more easily matched to conurbation size, fuel can be used in a wide range of plant sizes, lower capital costs.	Markets are not developed fully; markets need to be close to the plant and waste source to reduce transport costs; boiler operation needs careful attention to prevent boiler fouling; emission standards are not fully developed; suitable small-scale abatement systems need development on a commercial basis.
Pyrolysis and Gasification	Recovery of oil, a storable transportable fuel; potentially less polluting than full scale incineration or combustion; prospects for high efficiency power generation	Products can be contaminated requiring expensive cleaning; technology not fully developed; markets not developed.
Anaerobic Digestion		
—Engineered Process	Yields methane and a soil conditioner.	Technology not fully developed, potentially sensitive to feedstock composition.
—Landfill	Proven technology Low capital costs.	Requires long term maintenance and stricter safety controls. Sites limited by geology and geography.

preparation (especially for temporary roadways), or even for disposal to anyone wishing to collect the material. Likewise, the scope for selling or free collection from works depots of surplus broken paving slabs and other materials as a result of street maintenance works should be evaluated. Finally, it may be possible to screen soils and sub-soils for use as a growing material.

5.90 The re-use of these types of waste must be considered carefully to avoid the problems of contaminated land.

CHAPTER 6
Barriers to Recycling

Introduction

6.1 Since reclaimed materials are used as alternatives to primary raw materials, they must compete in terms of price, quality and availability. Scrap processing and marketing of the reclaimed materials are highly competitive operations which are sensitive to market forces. Their commercial viability depends on the existence of a sufficiently wide margin between collection costs and sales value of the reclaimed materials to cover processing costs and provide an adequate return on capital investment.

6.2 To maintain acceptable margins, reclaimers must take account of economic factors which are closely linked with technical aspects of the wastes. Important technical aspects to be considered include:

- the composition and physical characteristics of the waste, in particular the type and form of contaminant materials present;

- the availability of suitable technology and processing capacity for removing the contaminants and upgrading the waste into marketable products; and

- the degradation of potentially reusable materials during reclamation and reprocessing.

6.3 Economic factors that reclaimers must assess include the amounts of recoverable materials in the waste, the market demand and selling price of the reclaimed materials, the costs of collection, processing and marketing taking account of environmental and other regulatory controls, and the costs of disposal of the residues from the processing operations.

6.4 It is difficult to separate technical from economic factors in recycling operations because the market will determine the effort and investment that can be justified for the technology. Rising commodity prices will encourage greater technical input. Nonetheless, the two main barriers to recycling are the technical one of contamination and the economic one of market supply and demand. These barriers are discussed in the following sections.

Technical barriers

6.5 Reclaimed materials come from wastes which are rarely pure or clean enough for direct reuse. Household wastes are mixtures of potentially reclaimable materials and gross contamination—and there will also be minor contamination by dirt, grease, moisture, and other materials. Many of the contaminants can be removed efficiently by sorting, cleaning and refining processes, but there are limitations on the removal of some contaminants particularly if they are chemically or physically bound into the structure of the materials.

6.6 Manufacturers demand that their raw materials conform to specifications which strictly limit the nature and levels of contaminants. Although these specifications are often based on the levels of contaminants found in primary materials, it is necessary for recycled materials to meet them if they are to compete effectively. Because different applications require different standards, reclaimed materials will often complement primary materials in providing a range of products which meet all the market requirements. However, it is necessary either to remove the contaminants or to dilute them with material having low contaminant levels.

6.7   Contaminants in reclaimed materials fall into two categories:

i)   those which are not removed during pre-treatment and processing operations and which impair the quality of the recycled material or product—commonly referred to as residual contaminants; and

ii)   those which can be removed by processing but where removal reduces the yield of the reclaimed product, extends processing times to allow contaminants to be reduced to acceptable limits, or leads to the discharge of toxic fume, effluents or solid waste, which requires additional abatement measures—these are called non-residual contaminants.

6.8   Contaminants are a problem in most commonly recyclable materials. These are summarised in Table 6.1.

6.9   In **metals**, the presence of other metals can lead to micro-structural defects which cause brittleness, poor surface finish and cracking. For example, the residual contaminants (copper, tin and nickel) cause these defects in steel during hot rolling operations. Zinc is present in steels as a non-residual contaminant in the form of electro-deposited coatings and zinc phosphate primers. Although it is technically feasible to remove the zinc as a fume, the process requires longer refining times and creates greater waste disposal problems. Typical contaminants present in reclaimed aluminium scrap are shown in Table 6.1. The main problems are caused by alloys with other elements and other metals which are not removed before remelting of the scrap takes place.

6.10   Contaminants in **wastepaper** dictate the standard and the quality of the final product. As has already been described in Chapter 5, wastepaper is used as a substitute for primary pulp for writing, printing, wrappings and tissues and as bulk or packaging grades where the use of primary pulp is uneconomic. The most difficult contaminants are latex adhesives, plastics and more recently flexographic inks which are not removed by current flotation de-inking technologies and water-resistant coatings which prevent or slow down the pulping process.

Table 6.1 Potential Contaminants in Recycled Materials.

Recycled Material	Residual contaminants	Non-residual contaminants
Iron and Steel	Copper, tin, nickel	Zinc
Aluminium	Iron, silicon.	Lithium, glass, siliceous dirt, magnesium,zinc,tin, lead
Paper	Flexographic inks (>10%), water-resistant coatings	Adhesives, wire staples, plastics
Glass	Iron and chromium colourants	Metals, ceramics
Plastics	Fillers, colourants.	Other polymers, bacteria, inks, labels, adhesives
Compost	Heavy metals	glass

Note:-depends on physical & chemical state & concentration

The non-residual contaminants have the effect of reducing the quality of the final product, for example from a pulp substitute to a packaging grade.

6.11   Apart from adventitious contaminants, wastepaper is degraded by reprocessing which reduces the length of the fibres and hence the mechanical strength of the product. The presence of these degraded fibres is essentially as a contaminant which reduces the quality of the resultant paper and limits its range of application. The removal of the short fibres constitutes part of the shrinkage which occurs when paper is recycled.

6.12   For **plastics**, there is also a degradation of mechanical properties because of the high temperature moulding processes which reduce polymer chain lengths. Because these shortened polymers cannot be removed, various additives (eg impact modifiers) are used to improve mechanical properties.

6.13   Plastics derived from household waste usually consist of a mixture of different polymer types, which cannot be processed easily unless they are separated. This fundamental difficulty of mixing different polymers means that identification and separation systems must be used. Compatibilisers are available to process some polymer combinations, but the products tend to be low value products such as wood substitutes. Plastics from household waste are also contaminated with dirt, labels, printing inks and food residues which must be removed by cleaning processes. However, it is not usually possible to return recycled plastics to applications where good aesthetics are required (eg transparent film). Pigmenting has to be used to mask the presence of contaminants.

6.14   Apart from the aesthetic limitation, recycled plastics may not currently be used in direct contact with food and beverage products because of the difficulties of ensuring completely reliable sterilisation of bacterial contaminants and complete removal of chemical contaminants which may have diffused into the plastics. Some composite materials have little recycling potential apart from use as a fuel.

6.15   For **glass** aesthetics are the main property that is affected by the presence of contaminants. The colour of glass ie green or amber is dependent on the iron and chromium content and the chemical state of the metal ions in the glass. Different colours of glass cannot be reliably separated by automatic methods, and chemical methods for removing the metal ions from the glass are not available. As a result, colourless glass cannot be made from cullet contaminated with coloured glass, and although amber glass can be made from mixed amber and green cullet, the amount of green cullet acceptable in amber glass is limited. The net result is that most mixed cullet can only be used for green glass production, where the UK market is limited (15% of the total container glass market).

6.16   Metals are the contaminants of major concern in the application of **compost** derived from household waste. Heavy metals such as lead, cadmium and mercury originate from batteries, household dust and various household chemicals, and can become more concentrated by the composting process. Since little fundamental data is available on the uptake of heavy metals from composts, specifications of heavy metal concentrations for sewage sludge tend to be used as guidelines. Because no technology has been developed to remove heavy metals from compost, their presence as contaminants may limit the applications of composts to low value uses.

6.17   In summary, contaminants clearly present a barrier to recycling either by making the reclaimed materials unusable in a few cases, or more generally, to degrade properties and to limit the range of application to lower value products than for the primary materials. However, it should be remembered that there is a constant search for technical innovations to deal with these problems. This means that there is steady progress in many areas to overcome the difficulties of contaminants. This year's insuperable problem may have been solved in a few years time because of improved processes or better sorting and separation techniques.

Economic Barriers

6.18   Recycled materials must compete across a range of applications with primary materials. The recycled material will have to meet varying specifications, depending on the application. In practice, the presence of contaminants together with degradation of materials during re-processing means that impure recycled materials tend to be aimed at lower specification markets which are also lower value.

6.19   Demand for materials is dependent on the level of industrial activity which is clearly cyclical. As supply and demand change, the price of primary raw materials changes which, in turn, affects the price of recycled materials. For example, between 1985 and 1989, the price of aluminium scrap roughly doubled, and the price of wastepaper almost halved.

6.20   In assessing supply and demand, it is important to look beyond the UK. Many reclaimed materials are internationally traded commodities, in which an imbalance of supply and demand can lead to rapid changes in prices. The introduction of recycling schemes in other countries can lead to surpluses of materials which are then imported to the UK, thus depressing prices for recycled materials coming from within the UK. Conversely, exports of reclaimed materials provide a way of dealing with excess supply in the UK, a practice that has been adopted by the ferrous scrap industry for several years.

6.21   Over the past five years, annual collection of ferrous scrap has averaged about 9 million tonnes, whereas the UK iron and steel industry has used about 5.6 million tonnes per year. Because the capacity of the industry to use more scrap is limited by the technology and the plants available in the UK, the excess scrap is exported. Increased scrap use requires additional investment in plant, but investment decisions depend on the forecast demand for the finished product. Decisions cannot be based solely on the requirement to recycle more scrap. Additional material collected from recycling schemes has to compete with existing sources of scrap or else be exported.

6.22   Exports have also played a significant role in balancing supply and demand in the aluminium industry. Competition from overseas for high grade aluminium scrap has led to annual exports of over 90,000 tonnes. This has had the net effect of making it more difficult to recycle lower grade scrap, which requires dilution of contaminants by high grade material or by primary aluminium. In the UK, consumption of scrap by secondary smelters has dropped in the past few years, as the motor industry's demand for castings has decreased. However, the construction of a 50,000 tonnes/year plant at Warrington by British Alcan Aluminium will provide a market for reclaimed cans which will be converted to new can stock.

6.23   The economics of wastepaper recycling are very sensitive to general economic activity and the demand for packaging materials. For example, the level of advertising in newspapers and magazines changes considerably between recession and expansion of the economy and this has a significant effect on paper & board consumption. The resulting cyclical surges in wastepaper consumption tend to have a disrupting effect on collection systems. At times of high demand, merchants are prepared to collect small quantities of wastepaper from voluntary groups and other schemes, but when demand slackens, merchants cut back their collection rounds.

6.24 Old newspapers and magazines from households are recycled principally to produce newsprint, tissues and industrial towelling. Current consumption of this waste is at an annual rate of about 550,000 tonnes as a result of the build-up of production at recently commissioned de-inking plants. Output at these plants is close to their design capacity, so further increases in wastepaper supply need to be linked to new processing capacity. If household wastepaper is not segregated, it comes within the grade of Mixed Paper which is used for the production of packaging and building board; this grade is less valuable than that used for de-inking, and competes with Container Waste A which is generally of better quality. As a result of competing waste products, it is important to match the quality of wastepaper supplies to the market demands for the different grades of paper.

6.25 Glass recycling also provides an example where quality and markets inter-relate. As the residual colourants in the glass dictate that clear glass cannot be produced from coloured cullet, and amber glass is not tolerant to mixing with green glass, markets are dictated by the colour of glass collected. Whereas in 1990, the UK container industry produced (by weight) 69% clear glass, 16% green and 15% amber glass, the cullet collected from the general public and commercial sources was 27% clear, 36% green, 8% amber and 29% mixed (mainly green and amber). This difference in colour balance reflects the fact that the UK exports mainly clear filled containers (eg for spirits) but imports mainly green containers (eg for wines and lagers). It also highlights the fact that people are better at putting coloured bottles in banks than clear jars; and are not always careful about segregating clear & coloured glass properly. The net effect of the imbalance is that UK green glass production

already uses over 39% cullet, whereas clear glass production uses about 12% cullet and amber glass uses less than 10% cullet. As recycling rates rise, the market demand for green cullet will need to adjust either by export of cullet, or increased UK bottling in green containers.

6.26 For the plastics and compost sectors, markets for reclaimed materials are still relatively small, but economic barriers exist even here, because of the costs involved in upgrading properties to meet standards. For plastics, the aim is to isolate and recycle materials made from one polymer, because of the difficulties of processing mixed polymers and the relatively low value of the mixed ploymer products.

Impact of Barriers to Recycling on Local Authorities

6.27 For local authorities, these technical and economic barriers are reflected in the prices offered for reclaimed materials by merchants and processors. Owing to the heterogeneous composition of household waste, separate collection of recyclables from households ensures the highest recovery of clean recyclables, which will command the highest prices. However, experience to date in the UK indicates that collection costs for this type of scheme could be of the order of £50 to £150 per tonne of recyclable materials collected. This is the anticipated weight of recyclables from about 300 households, based on a weekly collection cycle. Whilst some savings can be anticipated from the reduced costs of collecting the residual wastes and improved efficiency based on growing experience of collection systems, the collection costs for recyclables are a significant additional cost which makes it vital to find markets to generate sales revenue.

6.28 The revenue obtainable for one tonne of recyclables depends on the proportions of the various material types that are present. Table 6.2

gives illustrative weights of materials from a kerbside collection scheme that might make up this tonne of recyclables, along with price estimates for each material, and the calculated revenue contribution.

6.29 Clearly glass and aluminium provide the main revenue sources and it would be ideal if stable markets and prices for these materials could be maintained. It is pertinent to point out that bottle-bank glass recycling has seen steady growth over the past ten years and that during this period the glass industry has consistently provided stable prices. Whilst it is possible in principle to stock-pile other reclaimed materials when prices drop, in practice deterioration of materials in storage and storage costs can rapidly outweigh any gains from future price rises.

6.30 Apart from the revenue from the sale of reclaimed materials, authorities may of course obtain recycling credits or pass them on to others involved in recycling. The value of credits should rise as a result of the Environmental Protection Act increasing landfill standards and costs. Clearly, given the additional costs of collection and the relatively low value of the materials collected, the size of the credit will be a major factor influencing the overall economic viability of the scheme. Guidance on analysing the costs for recycling plans is given in paragraphs 3.39–3.52.

Conclusion

6.31 The account given in this chapter of the different barriers to recycling is not intended to demonstrate that recycling cannot happen. Indeed, schemes throughout the country prove that it can happen. Certain schemes are already claiming to have achieved the Government's recycling target. Others believe that they can recycle considerably more than 50% of their household waste.

6.32 Barriers are described here to give an indication of the obstacles that need to be overcome. As more recycling takes place, the incentive for industry and others to find solutions will grow. Similarly, an increase in the amount of reclaimed material potentially available through recycling collections will serve to make investment in recycling plants more attractive. One constraint on investment is uncertainty about whether the reclaimed material will be available in sufficient quantities and from sources whose reliability can be guaranteed in the long-term.

6.33 Achieving the Government's target of recycling 25% of household waste by the end of the decade will require a transformation of attitudes and practices, on the part of industry, collection and disposal authorities and the public. But the rewards of making that transformation could be immense.

Table 6.2 Revenue from Sale of Recyclable Materials.

Commodity	Weight distribution per tonne of recyclables (Kg)	Estimated Price (£/Tonne)	Revenue contribution per recyclables (£)
Paper and board	515	5	2.60
Plastics	70	25-75	1.75–5.30
Glass	280	20-30	5.60–8.40
Ferrous metal	120	20-30	2.40–3.60
Aluminium	15	400-700	6.00–10.50
Total	1000	–	21–39

Note: These prices relate to 1990 merchant prices, and provide no guide on future price trends. Up-to-date prices should be obtained from merchants and processors when costing recycling plans. The weight distribution is generalised from experience of existing source separation schemes.

Format of a Recycling Plan

Chapter 3 provides guidance on preparing a waste recycling plan. This appendix offers a model outline for the recycling plan. It may be useful to waste collection authorities for checking that they have covered all of the relevant points when drawing up their plans.

1. Introduction

— legislative background—Environmental Protection Act 1990
— other Government measures
— objectives of the Plan
— summary of targets and timetable (see also section 12 below)

2. Scope of the Plan

— geographical area
— description of households and other properties serviced
— wastes and materials included in the Plan

3. Waste Inventory

— quantities of waste generated
　— door to door collection
　— 'civic amenity' sites
　— recycling collection
— composition of wastes, for
　— household
　— commercial and other

4. Current Waste Collection and Disposal Practices

— collection methods
　— refuse collection system
　— 'civic amenity' sites
　— collections by voluntary groups and others
— waste disposal methods
　— transfer system
　— waste treatment system
　— landfill
　— incineration
— recycling collection systems
　— bring, collect and central sorting systems
　— separation and processing facilities

— markets for reclaimed materials/energy

This section should provide information and statistics on the proportion of each material using the different 'disposal' routes (landfill, materials recycling, composting, etc). It should also indicate the main organisations involved in each stage of the process and their specific role.

5.   Future Waste Trends

— kinds and quantities of controlled waste which the authority expects to collect and purchase during the plan period
— kinds and quantities of controlled waste which the authority expects to recycle during the plan period

6.   Market Opportunities for reclaimed Materials and Energy

— national and local markets for
  — paper
  — glass
  — ferrous metals
  — aluminium
  — plastics
  — compost
  — energy as heat, power or fuel
  — others
— present situation and future trends in supply and demand. How variable are supply and demand for each material and what are the authority's plans for catering for fluctuations?
— authority's corporate purchasing policy

7.   Collection Systems to Meet Market Requirements

— review of bring, collect and central processing systems in relation to the local situation, to meet market requirements and the Government recycling targets
— design of the collection system

8.   Processing Systems to Meet Market Requirements

— central processing facilities and/or equipment
— role of the Waste Disposal Authority, waste disposal contractors and other key organisations eg scrap reclamation and materials processing companies

9.   Financial Analysis of Recycling Options

— financial spreadsheet taking account of
  — discount rate
  — capital costs
— equipment operating costs

- equipment operating costs
- transport costs
- labour costs
- administrative costs
- promotional costs
- disposal costs, including costs of after-care
- any other relevant costs
- revenues from sale of reclaimed materials
- revenue from commercial sponsorship

any other relevant revenue

estimated costs and savings from implementing the Plan

The analysis needs to cover the existing situation, the transition phase when recycling schemes are being introduced, and the state when recycling schemes are operational.

10. Plan for the Development of Recycling

- statement of the arrangements which the authority has made or proposes to make for recycling waste, or that others have made or propose to make, in relation to
  - materials to be reclaimed
  - collection and separation methods
  - processing facilities
  - marketing arrangements

  The reasons for selecting particular options should be clearly set out.

11. Consultation, Publicity and Education

- consultation plans
- the recycling message
- audiences to be targeted
- methods of publicising and promoting the Plan
- education programmes
- estimated sponsorship and costs

12. Implementing the Plan

- introducing the Plan
  - installing equipment and facilities
  - negotiating of market contracts
  - practices of waste collection staff
  - practices of householders
  - managing and policing the Plan
- targets and timetable for implementation
  - monitoring and assessment
  - revising and updating the Plan

APPENDIX B

UK Statistics on Recycled Material

Table B.1 Wastepaper Recovery/Collection in the UK
(Quantities in thousands of tonnes)

	1983	1988	1989	1990
Paper and Board Production	3,298	4,292	4,495	4,824
Wastepaper used	1,827	2,417	2,578	2,847
Utilization (%) *	55.4	56.3	57.4	59.0
Total P & B Consumption by end-users	7,159	9,283	9,599	9,273
Wastepaper recovery **	1,992	2,781	2,975	3,092
Recovery (%) †	27.8	30.0	31.0	33.3

Source:— BPBIF

*   Utilization—Wastepaper consumption as a proportion of production of total paper and board production.

†   Recovery—Wastepaper recovered as a proportion of total paper and board (apparent) consumption. (Adjusted for changes in stock levels).

**   Includes net exports.

Table B.2 Consumption of Wastepaper in UK 1985–1990
(Quantities in thousands of tonnes)

Waste Grades	1985	1989	1990
Group 1	26.7	48.5	55.4
2	80.4	50.3	60.7
3	18.9	74.8	73.1
4	139.5	146.3	185.9
5	259.9	449.9	556.5
6	30.3	31.5	28.2
7	275.9	269.5	260.4
8	646.4	927.5	1036.3
9	550.1	502.6	500.5
10	39.0	60.6	64.3
11	–	24.6	25.5
Total	**2067**	**2586**	**2847**

Source:— BPBIF

Waste grades are listed and defined in Appendix C.

Table B.3 Estimated UK Wastepaper and Board Recovery 1990

Type of Waste	Consumption	imports	Exports	Gross Weight Collected
Group 1	55.4	8.1	26.9	74.2
2	60.7	10.3	34.4	84.8
3	73.1	69.2	30.2	34.1
4	185.9	11.6	111.6*	285.9
5	556.5	10.7	63.1	608.8
6	28.2	0.6	1.4	29.0
7	260.4	5.8	13.3	267.9
8	1036.3	24.0	55.4	1067.7
9	500.5	12.8	14.1	501.8
10	64.3	4.0	38.6	98.9
11	25.5	1.6	15.4	39.2
Totals	2846.6	158.7	404.4	3092.3

Source:— BPBIF

*May include Group 9 and Group 10 materials.

Table B.4 Ferrous Scrap Utilisation in the UK.
(approximate ferrous metal derived from scrap)

Iron and steel making,	
by Basic Oxygen Systems	25%
by electric arc systems	100%
Iron castings	88–90%
Steel castings	100%

Table B.5 Use and Recovery of Ferrous Scrap in the UK.

	1985	1989
UK Home Consumption	11,615	14,904
Ingot Steel Production (A)	15,722	18,740
Iron Casting Production (B)	1,209	1,121
Total Scrap Used (C)	8,396	8,996
Scrap From Steelworks and Foundries (D)	3,318	3,227
Net Bought in Scrap* (E)	5,070	5,910
Scrap imports (F)	51	85
Scrap Exports (G)	4,520	3,250
Apparent Collection (H=E+G–F)	9,539	9,075
Utilization Rate (I=C/(A + B) %)	50	45
Scrap Use in Steel Making (J)**	6,994	7,853
Average Scrap Consumption in Steel Making (K = J/A)	44.5	41.9

Source:— Warren Spring laboratory estimates

All Quantities are in thousands of tonnes.

Adjusted for scrap stock changes and includes steel work scrap eg from old plant, ingot moulds

Table B.6 Non-Ferrous Metals in Household Waste

Metal	Form	Modes of Occurrence
Aluminium (Al)	wrought alloy	Cans, cooking and wrapping foil, pie and other containers, bottle closures, domestic hollow-ware, door furniture (handles etc), PVC coated extruded metal, anodised decorative articles, sub-components and wire in domestic appliances.
	cast alloy	sub-components in domestic appliances, door locks, door and furniture handles.
Copper (Cu)	metal	insulated and uninsulated wire, electrical fittings, components in electrical goods, kitchen ware and hollow-ware decorative goods, plumbing fittings.
	alloy	as a constituent of brass (used as above), screws, hooks and other fasteners, as a constituent of the cupro-nickel substrate in silver and other plated products.
	electroplate	electroplated ferrous tube lron heating systems, decorative waste.
Zinc (Zn)	metal	carbon-zinc batteries, battery cases.
	alloy	as brass:—see copper above, as aluminium casting alloys—see above, as diecastings, door fittings, handles, components in domestic and garden appliances, toys.
	galvanised	domestic kitchen and garden ware.
Tin (Sn)	electroplated	cans, boxes and other containers, kitchen ware, toys and sub-assemblies in household appliances, tinned Cu wire and electrical contacts.
	alloys	solder (electrical and plumbing joints), as a constituent of bronzes and some copper alloys, bearings in electromechanical appliances.
Lead (Pb)	metal	pipes, wine bottle closure capsules, electric bulb contacts.
	alloys	grids in lead-acid batteries, solder (plumbing and electrical joints) bearings in electromechanical appliances.
	chemical	as oxide and sulphate sludge in lead acid batteries— as a constituent of lead-based paints, eg red lead primer.
Nickel (Ni)	metal	as plating on numerous articles
	alloys	cutlery and other stainless steel articles, components of household appliances (washing machine drums).

Table B.7 Recycling of Aluminium in the UK.

	1985	1990
Primary Production (A)	275	290
Secondary Production—Gross Ingot Output (B)	128	121
Primary Al used in Secondary Production (C)	5.7	6.6
Net Scrap used in Secondary Production (D=B–C)	122	114
Total Production (E=A+D)	397	404
Consumption of Primary (F)	350	441
Consumption of Secondary and Direct Use (G)	109	81
Total Consumption (H=F+G)	459	522
Secondary Metal Content (I=G/H)	22.4	16
Scrap used directly (J)	28.8	32
imported Scrap Consumed (K)	4.4	16.9
Exported Scrap(L)	80	94
Estimated Net Scrap Collected (M = D+J+L–K)	226	223
Collected Scrap Exported %	34	42

All quantities in thousands of tonnes
Source:—Compiled by WSL based on data from World Metal Statistics.
Primary Consumption is based on shipments to domestic customers.
Secondary Production excludes primary metal content and melting loss.
Aluminium consumption excludes net imports of semi-manufactured goods.

UK Standard Groups of Waste Paper

Group 1 (White Woodfree Unprinted)

Best White Shavings No 1
Best White Shavings No 2
White and Cream Envelope Cuttings
White Coated
White Printers Shavings
White Soft Tissue

Group 2 (White Woodfree Printed)

Best One-cuts
Black and White Best Pams
Book Quire
Sulphate Waste
Tear White Shavings
White Carbonless Copy Paper NCR)
White Continuous Business Form
White Listings No I (CPO)
White Listings No 2 (CPO)

Group 3 (White and Lightly Printed Mechanical

Lightly Printed Scanboard
White Mechanical Coated
White Mechanical Listings (CPO)
White Scanboard (Duplex)
White Unprinted News
Woody One-cuts

Group 4 (Coloured Woodfree)

Best Pams
Coloured Carbonless Copy Paper NCR)
Colour Continuous Business Forms
Coloured Heavy Letter
Coloured Mill Broke
Coloured Shavings
Coloured Tissue

Multigrade
Sulphite Bag Waste
White and Light Toned
Shavings

Group 5 (Heavily Printed Mechanical)

Crushed News
Green Mechanical Listings
Heavily Printed Scanboard
Mechanical Book Quire
News and Pams
Over Issue News and Pams
Telephone Quire

Group 6 (Coloured Crafts and Manilas)

Buff and Coloured Tab Cards
Buff Envelope Cuttings
Dark and Light Coloured Manilas
Kraft Liner
Multiple Kraft Sacks
New Brown Krafts
Old Brown and Coloured Krafts

Group 7 (New Kraft Lined – KLS)

Double Lined Kraft (DLK)
New KLS Cuttings

Group 8 (Container Waste)

Container Waste (Old LKS)

Group 9 (Mixed Papers)

Mixed Papers

Group 10 (Coloured Card)

Coloured Card

Group 11 (Contamintaed Grades)

Wet-strength Paper and Board
Woodfree Paper Mixtures
Plastic Coated Paper and Board
Bookbinders Cuttings (Latex Tips and
Bindings)
Laminated Waste
Telephone Directories

Groups 1-4 are pulp substitute grades
Groups 6-10 are packaging/bulk grades

APPENDIX D

Summary of Energy Recovery Options

Direct Incineration

D.1   Incineration of municipal waste started on a large scale in the UK in the late 1960s although its use for energy recovery can be traced back to the late nineteenth century. Currently there are over 30 municipal waste incinerators in the UK of which 4 have energy recovery systems. These plants treat less than 10% of the UK municipal solid waste arisings. Modern continuous incineration plants receive household, commercial and selected industrial waste. It is burned without pre-treatment and produces heat, much of which can be recovered by suitable boiler plant. Steam so produced can be used to generate electricity. The most economic arrangement therefore requires the plant to be close to the waste source and also to customers for the heat or energy produced.

D.2   Operationally, the main problems of incineration arise from the extremely variable feedstock which gives rise to fluctuating emissions. Incinerators now require sophisticated pollution abatement techniques to deal with the pollutants in the waste especially as there is seldom pre-treatment of the feedstock. Such techniques demand equipment which is expensive to install and requires expertise to operate. Generally only large arisings—typically in excess of 100,000 tonnes/year—are considered adequate to justify the capital cost and operating costs involved.

D.3   Pollutant emissions arise either from material present in the waste to begin with, or from material generated by the combustion process. Household waste contains materials such as lead from lead-based paints, mercury

and cadmium from batteries and chlorine from paper and plastics. Incineration at high temperatures (>800°C) leads to the release of these elements, and also the release of other pollutants including oxides of sulphur and nitrogen, hydrogen chloride, and various organic compounds either present in the waste or formed by incomplete combustion. The potential formation of dioxins has been studied intensively to ensure that combustion and abatement techniques reduce emissions to negligible proportions. Nevertheless, the problem has not been entirely satisfactorily dealt with on many plants; mainly because of lack of instrumentation to control the combustion conditions adequately and the generally poor mechanical state of some of the older grate systems.

D.4   The control of pollutant emissions requires the use of electrostatic precipitators or bag filters to remove dust and grit, and the use of scrubbers to remove acid gases. This equipment will be essential for all new plants to meet European Community Directive (89/369/EC). Trapped material and residual ash must also be treated with care, as they may retain heavy metals and other toxic materials.

D.5   Apart from mixed household waste, incineration has been used for the recovery of energy from a range of agricultural and other industrial wastes. For example, a small unit capable of burning 100 tyres/hour has operated at Avon Tyres in Wiltshire for several years. In late 1990, Elm Energy and Recycling received planning permission to build a new incinerator to burn up to 100,000 tonnes/year at Wolverhampton. The plant is designed to have a power output of 20MW. Other companies have put forward plans to build a further 3 small incinerators, with energy recovery. Used tyres will come from tyre distributors, local authorities and from vehicle dismantlers and fragmentiser operators. In a unique example of an integrated waste management solution, Clackmannan District Council operates a combined refuse and sewage sludge disposal facility in which pulverised household waste is used as a fuel to dry and incinerate sewage sludge.

Refuse Derived Fuel (RDF)

D.6   Solid RDF falls into two main categories:

i   non-densified, i.e. shredded or fluff.

ii.   densified, i.e. pelletised or briquetted.

Non-densified RDF can vary from shredded refuse which has undergone minimal processing, to an upgraded product consisting mainly of higher calorific value materials such as paper, plastic and wood. Densified RDF requires more extensive processing to produce a fuel which can be easily stored and transported for use as a coal substitute.

D.7   Systems for the recovery of non-densified RDF from refuse are generally based on a combination of size reduction, screening, classification and magnetic separation techniques. Several of these systems have been built in the USA. In the UK, one of the simplest approaches to the use of shredded refuse waste was to shred the material to less than 70–80mm in size and after magnetic removal of ferrous material, the product was injected into a moving grate coal-fired furnace where combustion took place above and on the bed.

D.8   An alternative use for crude, undensified RDF is in admixture with coal in cement manufacture. Although doubts have been expressed in the US as to the economic and technical feasibility of converting cement kilns to utilize RDF, a process was developed by Blue Circle Industries for firing their wet process kilns at Westbury with waste supplied by Wiltshire County Council. The amount of RDF burned, however, is constrained by the amount of refuse ash which can be absorbed by the cement. The company claim savings of 12–13% of the primary fuel normally consumed.

D.9   The general use of shredded RDF as a substitute for or additive to coal is constrained by the limited number of commercial installations where it can be burnt efficiently. However densification of the RDF by pelletisation or briquetting decreases transport costs and improves its handling and combustion characteristics for direct substitution for coal in existing systems. Densified RDF is generally produced from an air classifier light fraction via further processing stages of milling, drying and pelletising or briquetting. A typical plant is operating at Byker in Newcastle, where annual production is 12,000–15,000 tonnes RDF.

Pyrolysis and Gasification

D.10   Pyrolysis is the thermal decomposition of materials in an inert atmosphere, to produce a fuel mixture consisting of gases, liquids and a solid char. Typically, operating temperatures are 70°C. Pyrolysis is seen as a potential alternative to incineration of wastes because a storable and transportable liquid fuel can be produced which is potentially easier to utilise. The high capital costs associated with pollution abatement for incinerators may also be obviated. However, because the high cellulose content of the feedstock yields a relatively low calorific value material, this potential has not yet been realised in practice. On the other hand, the application of pyrolysis to energy recovery from higher calorific value materials, particularly used tyres and plastics may be more feasible. Experimental work on tyres and plastics has taken place in several countries, but there are no plants which have yet demonstrated commercial success.

D.11 Research is now being directed towards advanced (flash) pyrolysis for the recovery of higher value liquid products, particularly from wood and other biomass feedstocks.

D.12 Pyrolysis has not yet been adopted by any UK local authority for waste disposal but a variety of pyrolysing systems have been examined as possible methods for energy recovery from refuse in overseas countries.

D.13 Gasification is the thermal destruction of materials in the presence of a limited amount of air—much less than would be necessary for complete combustion of materials. The products of the process are fuel gases—either low calorific value (5–6 MJ/m3) or medium calorific value (15 MJ/m3) synthesis gas depending on the process conditions. The process is fairly well established in North America for some fuels such as wood-waste, but has not been introduced in the UK. The main advantage of gasification is that the process can cope with a wider range of wastes than full combustion or pyrolysis. In addition, gasification makes it easier to retain pollutants in the ash instead of being generated in the gas phase and being discharged to the atmosphere.

Anaerobic Digestion with Methane Production

D.14 The biodegradation of refuse using anaerobic bacteria to produce methane has been demonstrated over a number of years. The process can be integrated closely with sewage disposal with possible significant savings in capital requirements. Traditionally the process comprises the fermentation of mixtures of macerated refuse and sewage sludge in large anaerobic digestion tanks. Initially the organic matter is broken down by bacteria into simple and soluble materials (organic acids) which in turn are converted by methanogenic bacteria to

methane. In practice about 50% of the decomposed organic materials are converted to methane. Estimates which take account of the energy required to maintain the digester temperature suggest that recovery of 25% of the potential intrinsic energy content of refuse is possible.

D.15 Two main approaches are being investigated:

- low solids systems where pulp densities of less than 10% weight solids are used ie similar to conventional sewage sludge anaerobic digestion plant; and

- high solids systems where solids contents of up to 40% are digested.

D.16 The latter approach greatly reduces required digester volumes and minimises subsequent water treatment and disposal problems, but the technique is less well developed commercially. Although full-scale plant of both design types have been demonstrated abroad for household waste feedstocks, to date only pilot or prototype test systems have been studied in the UK. In one example using a low-solids method, a wide range of household and commercial waste was pulverised, screened to remove non-pulpables such as metals and plastics, before mixing with sewage sludge and fermenting in a digester. Methane was recovered for use as a fuel and the residue was subsequently processed into three products, a solid fuel, a fibre product for board making or a soil conditioner.

D.17 Engineered anaerobic digestion is seen as a potential growth technology for the future. It is compatible with the sorting technology used to produce RDF and with source separation systems. In addition it is one of a number of technologies where research is receiving Government support. For example, Leeds City Council (in collaboration with the Department of Energy and the private sector) are developing a trial anaerobic digester for the separated organic fraction of household waste which should produce methane more rapidly and enable the digestate to be composted with further aerobic treatment.

D.18 Anaerobic digestion of wastes in landfill sites coupled to collection of the gases has a significant role in energy recovery from household waste in the UK. Landfill gas contains methane and carbon dioxide in varying percentages together with hydrogen sulphide and other minor constituents and it is saturated with water. Landfill gas is being exploited as a source of energy in the UK and several European countries. In general, wells are sunk into the landfill using perforated pipes, vacuum applied to the collection system and the gas is then piped to a processing plant. The gas can be suitable for direct use but processing may be required to remove water and particulates. Further upgrading by gas-cleaning is possible.

D.19 In mid 1990, landfill gas projects in England and Wales contributed electricity generating capacity of 17.8MW, with new projects planned to contribute another 25MW. Other landfills are supplying gas to boilers, kilns and a leachate treatment plant. Landfill gas can migrate through soils and can inhibit plant growth, therefore careful engineering of the site is essential. In addition, the hazards of fire and explosion must be eliminated by sealing of the surfaces. The containment and sealing of landfill sites, together with the practices for landfill gas extraction are the subject of Waste Management Paper No 27.

Calorific values of waste components

D.20 The calorific values of individual components of the waste are listed in Table D.1. The values range from as low as about 4 MJ/kg for some food and garden waste to about 30 MJ/Kg for some plastic waste materials. These values are typical of these components after removal from mixed household waste. These components are contaminated by associated moisture and dirt. There is less of this contamination in separately collected waste. As a result, the calorific values will increase in these components to values closer to those obtained from clean, dry samples. Even so, it must be appreciated that the materials retain some moisture when apparently dry and many components have non-combustible components associated with the base material eg metal zips on clothing, staples in paper, metal foil in paper and plastic laminates.

Table D.1 Typical Heating Value of Principal Constituents of Household Waste

Materials	Calorific Value (CV) MJKg	Moisture Content (Wt %)
	As Received	As Received
Paper and paper products	12	30
Plastic film	27	25
Dense plastics	30	15
Textiles	15	25
Miscellaneous combustible	13	25
Putrescibles	6	65
Fines (<10mm)	4	40

APPENDIX E

Sources of Information on Recycling

Alliance for Beverage Cartons and the Environment, 24-28 Bloomsbury
Way, London, WC1A 2PX.
Telephone: 071 831 6262

Alcan Aluminium Can Recycling, 3rd Floor, Eldon House, Regent Centre,
Gosforth, Newcastle upon Tyne, NE3 3PW.
Telephone: 0670 813811

Alcoa Manufacturing (GB) Ltd, PO Box 68, Swansea, SA1 1XH
Telephone:0792 873301

Aluminium Can Recycling Association, Suite 308, I-Mex House, 52
Blucher St, Birmingham, B1 1QU.
Telephone: 021 633 4656

Aluminium Foil Recycling Campaign, 38-42 High Street, Bidford-on-Avon,
Worcs, M50 4AA
Telephone: 0789 490609

Association of British Reclaimed Rubber Manufacturers, First Avenue,
Trafford Park, Manchester, M17 1DT
Telephone:061 872 1424

Association of County Councils, Eaton House, 66 Eaton Square, London,
SW1W 9BH
Telephone: 071 235 8458

Association of District Councils, 26 Chapter Street, London SW1P 4ND
Telephone:071 233 6868

Association of Recycled Paper Suppliers, Bow Triangle Business Centre,
Unit 2, Eleanor St, London E3 4NP
Telephone: 081 980 2233

Association of Metropolitan Authorities, 35 Great Smith Street, London,
SW1P 3BJ
Telephone: 071 222 8100

British Carton Association, 11 Bedford Row, London, WC1R 4DX
Telephone: 071 242 6904

British Glass Manufacturers Confederation, Northumberland Rd, Sheffield,
S10 2UA
Telephone: 0742 686201

British Paper and Board Industry Federation Ltd, Papermakers House, Rivenhall Rd, Swindon, SN5 7BE (includes the Pulp and Paper Information Centre)
Telephone: 0793 886086

British Plastics Federation, 5 Belgrave Sq, London, SW1X 8PH
Telephone: 071 235 9483

British Scrap Federation, 16 High St, Brampton, Huntingdon, Cambs, PE18 8TU
Telephone: 0480 455249 (Umbrella body for ferrous metals reclamation industries)

British Secondary Metals Association, 25 Park Rd, Runcom, Cheshire, WA7 4SS
Telephone: 09285 72400 (Non-ferrous metals)

British Textile By-Products Association, Thorncliffe, 115 Windsor Rd, Oldham, Lancs, OL8 1RQ
Telephone: 061 624 3611

British Waste Paper Association, Alexander House Business Centre, Station Rd, Aldershot, Hants, GU11 1BQ
Telephone: 0252 344454

Chemical Industries Association, Kings Buildings, Smith Square, London SW1P 371
Telephone: 071 834 3399

Chemical Recovery Association, 9 Larch Grove, Paddock Wood, Tonbridge, Kent TN12 6LA
Telephone: 0892 833587 (Solvents & waste oil recovery)

Combined Heat & Power Association, 3rd Floor, 35-37 Grosvenor Gardens, London, SW1W 0BS
Telephone: 071 828 4077

Confederation of British Industry, Centre Point, 103 New Oxford Street, London WC1A 1DU
Telephone: 071 379 7400

Convention of Scottish Local Authorities, Rosebery House, 9 Haymarket Terrace, Edinburgh EH12 5HZ
Telephone: 031 346 1222

Department of Energy, Energy Technology Support Unit, B156 Harwell Laboratory, AEA Technology, Oxfordshire, OX11 0RA
Telephone: 0235 821000 (Technical advice on use of waste to produce energy)

Department of the Environment for N. Ireland, Calvert House, 23 Castle Place, Belfast, BT1 1FY
Telephone: 0232 230560

Department of the Environment, Local Environment Quality Division, Room A222, Romney House, 43, Marsham St, London, SW1P 3PY
Telephone: 071 276 3000

Department of Trade and Industry, Environment Unit, 3rd Floor, 151 Buckingham Palace Rd, London SW1W 9SS
Telephone: 071 215 5000

Environment Resource Group, Centre for Human Ecology, Edinburgh University, 15 Buccleuch Place, Edinburgh EH8 9LN
Telephone: 031 650 3463

European Recovery & Recycling Association (ERRA), 83 Avenue E Mounier, B-1200 Brussels, Belgium (international companies concerned with multi-materials recycling)

Friends of the Earth, UK2000 Projects Unit, 26-28 Underwood St, London, N1 7JQ
Telephone: 071 490 1555

Friends of the Earth Scotland, Bonnington Mill, 72 Newhaven Rd, Edinburgh EH6 5QG
Telephone: 031 544 9977

Henry Doubleday Association, National Association for Organic Gardening, Ryton on Dunsmore, Coventry CV8 3LG
Telephone: 0203 303517 (advice on home composting)

Independent Waste Paper Processors Association, 25 High St, Daventry, Northants, NN11 4BG
Telephone: 03272 703223

Industry Council for Packaging and the Environment, Premier House, 10 Greycoat Place, London, SW1P 1SB
Telephone: 071 222 8866

Institute of Wastes Management, 9 Saxon Court, St Peter's Gardens, Northampton, NN1 1SX
Telephone: 0604 20426

Liquid Food Carton Manufacturers Association, 30b Wimpole St., London, W1M 8AA
Telephone: 071 486 3337

Local Authority Recycling Advisory Committee, c/o London Waste Regulation Authority, Hampton House, 20, Albert Embankment, London SE1 9LL
Telephone: 071 587 3000

National Association of Waste Disposal Contractors, Mountbarrow House, 6/20 Elizabeth St, London, SW1W 9RB
Telephone: 071 824 8882

National Tyre Recycling Association, The Meadows, Ryleys Lane, Alderley Edge, Cheshire, SK9 7UU
Telephone: 0625 582346

Northern Ireland 2000, Armagh House, Ormeau Avenue, Belfast, BT2 3HB
Telephone: 0232 238532

Oxfam, 274 Banbury Road, Oxford OX2 7DZ (Textiles)

Reclamation Association, 16 High St, Brampton, Huntingdon, Cambs, PE18 8TU
Telephone: 0480 455249 (Textiles)

RECOUP - Recycling of Used Plastic Containers, Manor Barn, Polebrook, Mr Oundle, Peterborough, PE8 5LN
Telephone: 0832 274759

Retread Manufacturers Association, Brampton Bridge House, 10, Queen St, Newcastle under Lyme, Staffs, ST5 1ED
Telephone: 0782 661944

Save-a-Can, Elm Lane, 19 Elmshott Lane, Chippenham, Slough, Berks, SL1 5QS
Telephone: 0628 666658

Scottish Enterprise, Energy and Environmental Technologies Group, 120 Bothwell Street, Glasgow G2 7JP
Telephone: 041 248 2700 (Advice and project support for recycling in Scotland)

Scottish Office Environment Department, Environmental Protection Division, 27 Perth Street, Edinburgh, EH3 5RB
Telephone: 031 244 3148

Second Life Plastics, Anaplast Ltd., Lundholm Rd, Ardeer, Stevenston, Ayrshire KA20 3NQ
Telephone: 0294 605111 (Agricultural plastics)

Steel Can Recycling Information Bureau, Kingsgate House, 536, Kings Rd, London, SW10 0TE
Telephone: 071 351 5208

Spastics Society, 12 Park Crescent, London W1N 4EQ (textiles)

Tidy Britain Group, The Pier, Wigan, WN3 4EX
Telephone: 0942 824620

UK2000 Scotland, 201 Drakemire Drive, Glasgow G45 9TD
Telephone: 041 634 2155 (Advice and support for community and voluntary recycling projects)

United Kingdom Reclamation Council, 16, High St, Brampton, Huntingdon, Cambs, PE18 8TU
Telephone: 0480 455249 (Umbrella body for reclamation industries and associated interests)

Warren Spring Laboratory, Recycling Advisory Unit, Gunnels Wood Rd, Stevenage, Herts, SG1 2BX
Telephone: 0438 741122

Waste Processing Association, The Runnymeade Malt House, Runnymeade Rd, Egham, Staines, Surrey, TW20 9BO
Telephone: 0784 34371 (Refuse-derived fuel)

Waste Watch, NCVO, 26 Bedford Sq, London, WC1B 3HU
Telephone: 071 636 4066 (Publications, policy, advice, especially for voluntary bodies)

Welsh Office, Industry Department, Cathays Park, Cardiff, South Glamorgan, CF1 3NQ.
Telephone: 0222 825111

Wiping Cloth Manufacturers Association, 31A Hill Avenue, Amersham, Bucks, HP6 5BX.
Telephone: 0240 34105

World Action for Recycling Materials and Energy from Rubbish (WARMER), 83 Mount Ephraim, Tunbridge Wells, Kent, TN4 8BS
Telephone: 0892 24626

APPENDIX F
Bibliography

1. *This Common Inheritance, Britain's Environmental Strategy (Gm 1200)*, HMSO 1990. ISBN 010 1120028

2. *Environmental Protection Act* 1990 HMSO 1990. ISBN 0105443905

3. *Waste sorting and RDF production in Europe.* J R Barton et al. Elsevier Applied Science 1985. ISBN 0853344078

4. *Design for recyclability.* M E Henstock. Institute of Metals 1988. ISBN-0 901462462

5. *Household waste. The business of reclamation.* County Surveyors' Society Report 4/31987.

6. 'Integrated Resource Recovery. Recycling from municipal refuse: a state of the art review and annotated bibliography.' S M Cointreau, C G Gunnerson, 1 M Huls and NN Seldman. *World Bank Technical Paper No 30* 1984. ISBN 082130447

7. *Metal resources and energy.* P F Chapman and F Roberts, Butterworths, 1983. ISBN 040810801

8. *Compost processes in waste management.* W Bidlingmaier and P l'Hermite, Commission of the European Communities 1989.

9. *Resource recovery guide.* J Abert, Van Nostrand Reinhold 1983. ISBN 0442202350

10. *Resource recovery.* A report of the Royal Commission on Environmental Pollution, HMSO 1985. ISBN 0117301329

11. *International overview on solid waste management.* L Bonomo and E Higginson, Academic Press 1988. ISBN 0121149757

12. *Resource recovery from municipal solid wastes.* Vol 1 and 2. L F Diaz, G M Savage and C G. Golueke, CRC Press Inc, 1982. ISBN 084935613 X (Volume 1); ISBN 0849356148 (Volume 2).

13. *Energy from biomass and wastes* XII D L Klass, The Institute of Gas Technology, Chicago, USA, 1988.

14. *Engineering for profit from waste.* Institution of Mechanical Engineering, PO Box 24, Northgate Av, Bury St Edmunds, Suffolk, 1988. ISBN 0852986734

15.   *Environmental impact of refuse generation in the UK*. M Woodfield, Warren Spring Laboratory Publications Sales, Gunnels Wood Rd, Stevenage, 1987. ISBN 0856244767

16.   *Review of MSW Incineration in the UK*. D Scott et al. Warren Spring Laboratory Publication Sales, Gunnels Wood Rd, Stevenage 1991. LR776(PA) ISBN 0856246344

17.   *Plastics recycling as a future business opportunity*. Plastics Institute of America, Stevens Institute of Technology, Castle Point Station, Hoboken, New Jersey, USA.

18.   *Recycling of waste paper in the UK - an overview*. R Tron, Warren Spring Laboratory Publications Sales, Gunnels Wood Rd, Stevenage, 1987. ISBN 0856244740

19.   *Plastics recycling in Europe*. R Derry, Warren Spring Laboratory Publication Sales, Gunnels Wood Rd, Stevenage, 1989. ISBN 0856245674

20.   *Plastics waste recycling in the UK - a position study*. R New. Warren Spring Laboratory Publication Sales, Gunnels Wood Rd, Stevenage, 1986. ISBN 0856244139

21.   *Contaminants in ferrous scrap*. R Tron, Warren Spring Laboratory Publication Sales, Gunnels Wood Rd, Stevenage, 1986. ISBN 0856243418

22.   *The laboratory characterisation of refuse derived fuel*. T W A Rampling and T J Hickey, Warren Spring Laboratory Publication Sales, 1987. LR643(MR) ISBN 085624502 X

23.   *Anaerobic digestion of industrial wastewaters and municipal solid wastes - a position study*. C B Saw, Warren Spring Laboratory Publication Sales, 1988. LR672(MR) ISBN 0856245321

24.   *Recycling Domestic Wastes in Scotland*, A Report to Scottish Enterprise, 120 Bothwell Street, Glasgow, G2 71P, Tom Read 1991.

25.   *Waste recycling: a regional perspective*. SERPLAN, 50-64 Broadway, London SW1.

26.   *Civic amenity waste disposal sites*. Luton College, Park Square, Luton, 1989.

27.   *Materials Reclamation Handbook 1990*. EMAP Vision Ltd, PO Box 109, Maclaren House, Scarbrook Rd, Croydon Surrey. ISBN 0956-2982

28.   *Recycling strategy for Ontario*. Recycling Advisory Committee, 119 King St W, 9th Floor, Hamilton, Ontario, Canada, 1989.

29. *Towards a Recycling Culture*, Confederation of British Industry, Centre Point, 103 New Oxford Street, London WC1A 1DU. 1991. ISBN 0852013477

30. *Recycle 90 (Plastics)*. Maack Business Services, Seestrasse 308, CH-8804 AU/ZH, Switzerland.

31. *Garbage solutions*. M Chertow, National Resource Recovery Association, US Conference of Mayors, 1620, Eye St NW, Washington DC 20006, USA, 1989.

32. *Comprehensive curbside recycling/collection costs and how to control them*. Glass Packaging Institute, 1801 K Street, NW Suite 1105-L, Washington DC 20006, USA, 1989.

33. *Community strategy for waste management*. Commission of the European Community, 200, Rue de la Loi, B1049 Brussels, Belgium, 1990.

34. EEC Directive 85/339 UK Data 19R6 - *A Report for Industry Council for Packaging & the Environment*. I. Boustead & G.F. Hancock.

35. *Environmental Practice in Local Government*. Association of County Councils, Association of District Councils & Association of Metropolitan Authorities, 1990.

36. *The Green Maze, Environmental Information and the Needs of the Public* ed. J Button, E C O Environmental Educational Trust, Avon Environment Centre, Junction Rd, Bristol, BS4 3JP, 1991.

37. *National Directory of Recycling Information.*
    Recycling: A practical guide for local groups. ISBN 0719913144
    Recycled Paper/Paper Recycling: a guide for office use. ISBN 0719912601
    A Harvest of Waste: A practical guide to re-use and recycling in Villages and Rural Areas.
    Getting the Credit
    from: Waste Watch, 26 Bedford Square, London, WC1B 3HU

38. *Sharing Resources: The case for collaboration between local authorities and the voluntary sector in promoting waste recycling.*
    Expanding Glass Recycling: A LARAC guide.
    from: LRRAC (Local Authority Recycling Advisory Committee) Hampton House, 20 Albert Embankment, London, SE1 1TJ.

39. *Recycling: The Way Forward.*
    Recycling Directory for your County.
    Setting up a can collection
    Setting up a bottle collection.
    Getting a bottle bank into the community.

How to set up a paper collection.
Paperchase: a guide to office waste paper collection.
How to set up Office Paper Collections.
A guide for Local Authority Recycling Officers (due Spring 1991).
from: Friends of the Earth, 26-28 Underwood Rd., London, N1 7JQ.

40. *Recycling Guides to Scotland* (by region)
from: Friends of the Earth Scotland.

41. *Economic Appraisal in Central Government* HM Treasury HMSO 1991 ISBN 0115600345

Up to date information and technical papers on recycling are available from journals including

- Materials Reclamation Weekly

- Waste Management Bulletin

- Resources, Conservation and Recycling

- ENDS Newsletter

- Scrap Processing and Recycling

- Warmer Bulletin

- BioCycle

- Journal of the Institute of Wastes Management

Printed in the United Kingdom for HMSO
Dd.293977, 6/91, C35, 3385/4, 5673, 155815